A New Exploration
of the
Canadian Arctic

Ronald E. Seavoy

hancock

house

ISBN 0-88839-522-1
Copyright © 2002 Ronald E. Seavoy

Cataloging in Publication Data

Seavoy, Ronald E
  A new exploration of the Canadian Arctic

  Includes bibliographical references.
  ISBN 0-88839-522-1

  1. Seavoy, Ronald E. 2. Prospecting—Northwest Territories.
3. Mines and mineral resources—Northwest Territories.
4. Northwest Territories—Description and travel. I. Title.
TN27.N6S42 2002   622'.1'097193   C2002-910757-1

Photographs: Ronald E. Seavoy
Maps: Ronald E. Seavoy
Line drawings: Brian G. Seavoy
Editor: Mary Greenshields
Production: Mary Greenshields, Ingrid Luters

*We acknowledge the financial support of the Government of Canada through
the Book Publishing Industry Development Program (BPIDP) for our
publishing activities.*

Published simultaneously in Canada and the United States by

HANCOCK HOUSE PUBLISHERS LTD.
19313 Zero Avenue, Surrey, B.C. V3S 9R9
(604) 538-1114 Fax (604) 538-2262

HANCOCK HOUSE PUBLISHERS
1431 Harrison Avenue, Blaine, WA 98230-5005
(604) 538-1114 Fax (604) 538-2262
Web Site: www.hancockhouse.com *email:* sales@hancockhouse.com

Contents

Thelon River Camp

Yellowknife

Additional books by the author are *The Origins of the American Business Corporation*, 1784-1855 (Greenwood, 1982); *Famine in Peasant Societies* (Greenwood, 1986); *Famine in East Africa: Food Production and Food Policies* (Greenwood, 1989); T*he American Peasantry: Southern Agricultural Labor and its Legacy*, 1850-1995 (Greenwood, 1998); *Subsistence and Economic Development* (Praeger, 2000); *Origins and Growth of the Global Economy*, 1450-2000 (Praeger, 2003).

Introduction

This book describes some little known arctic phenomena that are of interest to general readers. They are by-products of exploring the last North American frontier: the Arctic. This is not the traditional frontier associated with the limits of agriculture, but the new frontier of finding and producing industrial raw materials. Particularly, it is a search for metals. New mines of nickel, copper, lead, zinc, and uranium are essential for the prosperity of our urban industrial civilization because metals and energy are the sinews of industrial production. This book is about exploring for metal deposits in the Canadian Arctic.

A combination of favorable circumstances induced INCO (International Nickel Company) to undertake arctic exploration in the late 1950s and early 1960s. There was a rapidly increasing demand for metals, because the industries of Western Europe and Japan had recovered from the destruction of WWII, and the long term outlook for all metals was increased consumption with stable production costs and good profit margins. At the same time, the federal government implemented policies that encouraged arctic exploration. It was strongly in the national interest to discover new

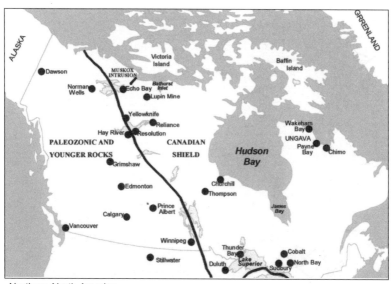

Northern North America

5

deposits of raw materials in the Arctic so they could contribute to national prosperity at a future date. In 1955, huge areas of the Arctic were blank spaces on the national map and it was in the national interest to extend the presence of government over this vast area.

The tax policies of the federal government encouraged mining companies to risk money on prospecting. Mining companies that undertook arctic exploration could writeoff 110 percent of the costs against taxable income. This was a substantial incentive for companies that had the technical and financial capabilities to conduct prospecting programs in remote places. These companies had skills that the federal government did not possess. In 1959, INCO undertook an extensive program of arctic exploration that lasted four years. The area of interest was the barren lands north of the treeline, as far north as the Arctic Ocean, and eastward from the eastern shore of Great Bear Lake to the Perry River.

In a reconnaissance investigation, during the summer of 1955, INCO sent its senior exploration geologist to the western Arctic. This investigation was an outgrowth of INCO's exploration program in northern Manitoba where extensive low grade nickel mineralization had been found. Late in the season, on high ground south of Coronation Gulf, he discovered a large rock formation that had many surface exposures of nickel mineralization. He named the formation Muskox Intrusion, because it was located where muskoxen still grazed. The Muskox Intrusion is located in the barren lands just north of the Arctic Circle.

INCO immediately undertook an intensive investigation to evaluate its potential for mineable deposits. By the end of 1958, diamond drilling confirmed that there were no mineable nickel orebodies associated with the Muskox Intrusion. Because of favorable tax incentives for arctic exploration, INCO decided to continue the program begun at Muskox. I joined the program at the beginning of the 1959 field season and remained until its end in September 1962. This book is an account of the events of 1960 and 1961, with a look back at the exploration of the Muskox Intrusion and a look forward to when the Lupin gold mine came into production in 1982.

An essential first step for opening the Arctic to mineral exploration was a systematic program of aerial photography in order to make accurate planimetric maps for airplane and helicopter navigation in this remote region. The federal government began this pro-

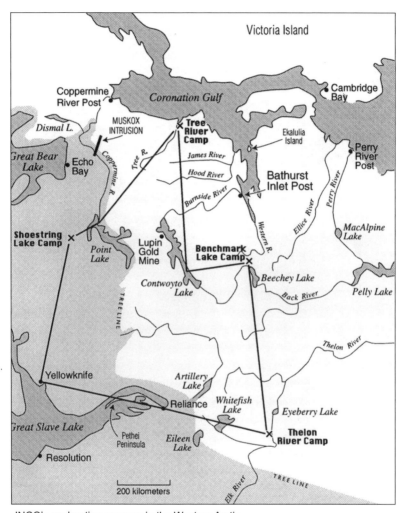

INCO's exploration program in the Western Arctic

gram immediately after WWII. At the same time, the Department of Transportation (D.O.T.) established interior weather stations that were sited to aid airplanes that supplied the DEW Line (Distant Early Warning) radar bases in the high Arctic. The radar stations were the first line of defense against a transpolar attack on North America. The weather stations were equally useful for ensuring the safety of exploration parties that could only be supplied by air.

In order to understand why the vacant spaces of the Arctic acquired industrial value in the 1950s, we must describe two developments: how the Arctic became accessible from the air and the development of exploration geophysics. Air accessibility was provided by float planes that had sufficiently powerful engines to carry freight on 350 kilometer round-trips, and helicopters that could land on rough terrain away from lakes. The De Havilland Beaver and Otter were the planes. The Beaver was designed in Canada, as a bush plane, and was first produced in 1949. This was INCO's principal supply plane. Bell made the helicopter. Its pilot and single passenger sat in a round plastic bubble cockpit giving both of them 230 degrees of visibility.

The essential supplement to accessibility and visibility was the application of geophysics to air and ground prospecting. INCO had tested these technologies in the discovery of the world-class nickel deposits at Thompson, Manitoba. In 1952, INCO discovered large low grade nickel deposits at Mystery and Moak Lakes in Manitoba. In 1955, INCO sank an exploration shaft at Moak Lake with the intention of preparing it for production. This new source of nickel would preserve INCO's free world leadership in nickel production. In 1955, INCO produced about 60 percent of the free world's nickel, and smelted and refined production from other mines, so that they handled about 70 percent of the nickel produced in the free world.

INCO's leadership was assured in January 1956 with the discovery of a high grade orebody located at the present townsite of Thompson, Manitoba. The Moak Lake project was abandoned and funds were transferred to the Thompson project. Definition drilling confirmed the existence of a large, high grade orebody. INCO's management made the decision to build a mine, mill, smelter, and refinery complex; a hydroelectric dam to supply electricity; and a new town to house workers. The complex became operational in March 1961 and continues to produce nickel in the year 2002.

The effectiveness of INCO's aerial and ground geophysical technology in locating the Thompson and satellite orebodies was an important ingredient in INCO's decision to spend 1 million dollars a year on arctic exploration. INCO had the personnel and technology that the government did not possess, and took risks that the government was not prepared to take. INCO's program was grassroots exploration of the most speculative variety. In the region that INCO would explore, nobody knew if favorable rocks existed because virtually no previous prospecting had been done.

The exploration program on the Muskox Intrusion operated out of Yellowknife on the north shore of Great Slave Lake. The town was built in the late 1930s when major gold deposits were discovered and brought into production. It is in the sub-Arctic south of the treeline, and is the northernmost city in Canada. Until 1961, it did not have a road connection to the outside world. Its only communication was a daily flight to Edmonton or a 200 kilometer boat trip from Hay River, on the south shore of Great Slave Lake, where the gravel road ended. Hay River was 520 kilometers north of the railhead at Grimshaw, Alberta. In the 40 years since 1960, the railroad has been extended to Pine Point, east of Hay River, to service major lead-zinc mines, and a paved road now connects Yellowknife with the outside world.

The economic potential of the Muskox Intrusion was so large and promising that INCO built a permanent camp of eight buildings. All of the lumber and other building materials arrived by chartered float planes. Muskox camp was built at the northern end of an unnamed lake that was six kilometers long and four kilometers wide. The first prospecting crews named it Desolation Lake because of its location in the barren lands. In the late summer of 1958, INCO began aerial prospecting by airplane flights radiating from Desolation Lake (now MacGregor Lake) whenever an airplane or helicopter could be spared from supplying drill camps or bringing supplies from Yellowknife. These flights began after the snow had melted and rock outcrops were exposed. At this latitude, runoff occurs late in June through to the first week in July.

In 1959, the aerial prospecting program was converted into a full-time operation. The new exploration program required two aircraft to be permanently attached to base camp: a Beaver to supply food and fuel and to move camp, and a helicopter for aerial

Northwestern Canada.

prospecting. The Beaver was also used to fly two-man prospecting teams to where geologically favorable rocks were located. Prospecting camps (fly-camps) were located on lakeshores, and two man teams prospected an area within walking distance. Base camp was moved progressively north, from Yellowknife as ice went off of the lakes. This meant that after prospecting personnel left Yellowknife in mid-June, both base camp and prospecting teams were always at temporary locations. Base camp was usually at one location for three weeks, but it could be longer or shorter depending on the weather. Fly camps were usually moved every week or ten days.

Both the Beaver and helicopter were equipped with floats. The Beaver was on floats because all camps were located on lakes with sand beaches. After the ice melted (breakup), all but the shallowest lakes could be used for landing. About 20 percent of the western Arctic is covered by lakes, thus float planes are an extremely flexible means of access. The helicopter was equipped with low pressure balloon floats, shaped like sausages, that allowed it to land on uneven ground. Floats were also survival insurance if the engine stalled over water. After a deadstick landing, the helicopter would float and could be paddled to the nearest shore, where repairs could be made, or where the pilot and geologist could wait until rescued.

Most of the Beaver's time was consumed with bringing supplies to base camp. Groceries came in every ten days, and three drums of aviation gas (a full planeload) were flown in every two or three days to keep the helicopter flying. When base camp was moved, an additional plane was chartered in Yellowknife to quickly complete the job. This caused minimal interruption in helicopter prospecting during the short summer season. When the plane was not delivering aviation gas or other supplies, or not moving two-man prospecting teams, it was used for aerial prospecting. The Beaver was especially useful in making longer flights to find a favorable site for the next base camp.

All of INCO's arctic exploration was on the Canadian Shield. The rocks of the Canadian Shield are the core of the North American continent, which has remained stable for the past 2 billion years. It is a vast area. Its easternmost provinces extend to the coast of Labrador; its southern boundary is along the St. Lawrence River, with a southward bulge that forms the Adirondack Mountains in northern New York. It then jogs north along the eastern and northern shores of Lake Huron to Sault Ste. Marie, Ontario, and again bulges south into central Wisconsin and northern Minnesota. From the northwestern corner of Minnesota, it turns north along the eastern shore of Lake Winnipeg. Thereafter, its boundary trends northwest to the south shore of the Great Slave Lake and, at the lake's center, it takes a more northerly direction to the Arctic Ocean.

The largest portion of the Canadian Shield is composed of granite, but it also contains many other kinds of rocks. Geologists search for the other rocks when prospecting from the air, because almost all mineral deposits are found in non-granitic rocks. The non-granitic rocks of the Canadian Shield have little relationship to the present climate and geography. The best evidence for this is the near absence of rocks containing evidence of life and, where evidence of life is found, it is algae of the same genera that live in tropical oceans today.

Most of the rocks of the Canadian Shield are highly deformed by large-scale folding and faulting that accompanied mountain building. During the earliest mountain building events, horizontal beds were compressed and folded into vertical positions and other rock units slid past one another for distances measured in tens to hundreds of kilometers. These same forces are still at work in mountain ranges along the western margins of North and South America and along the margins of other continents. But, for the past 2,000,000,000 years, these mountain building forces have been dormant in the Canadian Shield.

During most of this time, the rocks of the Canadian Shield have been emergent from the sea and mountain ranges have been eroded to their roots. In a few places, modest hills remain but, over most of the Shield, ancient mountain ranges are now low hills or flat plains that are 200 to 300 meters above sea level. During this prolonged period, there have been periods when portions of the Shield were buried by outpourings of basalt lava, or were flooded by a shallow sea and received a thin veneer of marine sediments. When sea level fell, most of these rocks were removed by erosion. The features that make the Canadian Shield a single geologic unit are its great age, structural complexity, and geologic stability.

The helicopter made it possible to prospect large areas of barren lands during the short arctic summer. In helicopter prospecting, the geologist sits in the passenger's seat and directs the pilot to land near mineralized outcrops so he can collect samples for assay. I was assigned the job of aerial prospecting for the 1960 field season. My job was to use my eyes to find mineralized outcrops and hope that they contained concentrations of metals of immediate or future economic interest. The possibilities were good that INCO would find a major mineral deposit. In the slang of exploration geologists, INCO hoped "to shoot an elephant."

My visual targets were gossans. Gossans are splotches of red that are caused by the weathering of iron sulfide minerals to iron oxide minerals. The most common iron sulfide mineral is pyrite (fool's gold). The second most common is pyrrhotite. Pyrite is found in small quantities in most rocks, but it is very abundant in deposits that contain concentrations of copper, lead, and zinc. Pyrrhotite is commonly associated with nickel mineralization. When concentrations of iron sulfide minerals are exposed to the atmosphere, the upper few centimeters oxidize. The most common iron oxide mineral is hematite. Its color is red. Therefore, concentrations of potentially valuable minerals advertise their presence by scarlet caps. In the barren lands, the red splotches of gossans are easily seen from the air, provided they are not covered by glacial till, hidden underneath lakes, or masked by the glare of the sun.

When I found a surface concentration of economic minerals, other technologies were used to evaluate the occurrence. For example, if I found an area with many small showings of copper sulfides but failed to find a large occurrence, INCO would bring in an airplane equipped with geophysical instruments. Aerial geophysics was a post-WWII development, and INCO was one of the first companies to use it on a large scale.

Aerial geophysical instruments send a strong electrical pulse into the ground. If there is a concealed deposit of sulfide minerals, receivers aboard the airplane record a response because sulfide minerals are highly conductive of electricity. A ground survey team then defines the location of the conductive zone with electromagnetic (EM) instruments, and geologists estimate its potential for having a base metal content by the size and intensity of the anomaly. Complementary information is gathered by a magnetometer survey that measures the anomaly's magnetic intensity. On the basis of this information, a decision is made to sample it by diamond drilling.

Most EM and magnetic anomalies are barren of base metal or precious metal mineralization.

This book takes the reader on a tour of the Canadian Arctic during the earliest years of the new exploration by airplanes, helicopters, and geophysics. The reader is an invisible member of INCO'S exploration team from the time I left Yellowknife in mid-June 1960 to when I returned in mid-September. In the course of the tour, I will describe and scientifically explain many little-known arctic phenomena.

CHAPTER 1

Tree Rings

The field season began on the June 15, 1960, when we flew from Wardair's dock at Yellowknife to our first base camp at Shoestring Lake, 275 kilometers to the north. Our Beaver, plus three chartered flights, supplied transportation. Our gear was piled on the dock for loading. It consisted of packs of folded tents tied with ropes, bundles of tubular aluminum tent frames wired together, geophysical instruments, a number of Coleman stoves, a large propane cookstove for base camp, and three space heaters. The tents were of two kinds: two small two-man tents for prospecting teams, and three large canvas tents for base camp. Base camp tents fitted over aluminum tube frames that snapped together in a few minutes, and they had metal doors in order to make them mosquito proof. Each tent rested on a platform built of plywood sheets nailed to two by four stringers. The tubular aluminum frames were securely nailed to the platforms, otherwise, high winds would blow them over. There was also a canoe, iron cots with mattresses, several boxes of cooking utensils, tools, rolls of stovepipes and wire, a keg of nails, a pile of duffel bags, a miscellaneous collection of fishing rods, and numerous boxes of food.

The first Beaver load was self-contained so that the two persons who unloaded it could stay and erect a tent while the plane was making a round trip. Within the plane, the plywood sheets, cots, and metal door were stacked against one side, and the rest of our gear was piled into the remaining space as best it would fit. The two by four stringers were lashed to the floats. I went with the first load and sat in the copilot's seat, and the student assistant sat on the keg of nails as far forward as it would go. He used a sleeping bag for a cushion.

The plane carried full tanks that made its load about 625 kilograms. It was overloaded by more than 130 kilograms because the rear of the floats were seven or eight centimeters under water, but we had a forty kilometer wind to aid takeoff. After we shoved off from the dock the plane wallowed among fifty to sixty centimeter waves until the engine was warm. Then, with full flaps and the engine turning 2,300 rpm, we began our run into the wind. As the plane gathered momentum, the pilot, Jim Shaver, jerked back on the stick causing the plane to climb on top of each wave and then drop back with a plowing lurch. Each time this happened the propeller would atomize the spray kicked up by the floats. He jerked back again and again until the plane was on the step and slapping the crests of each wave as it fought to become airborne.

Two or three efforts usually got the plane on the step and skimming the waves. If the plane is lightly loaded, the pilot tilts the plane to one side and the opposite float lifts from the water. The wing is then tipped the other way and the other float lifts from the water. This would not work today. Takeoff was a bull job, requiring full power to lift both floats in due time. The plane did not want to become airborne after it was on the step and it continued pounding the wave tops. The engine continued its full power roar for a longer time than I could remember, but gradually we began smacking the crests less heavily. Then we seemed airborne, but before I could be sure, the floats slapped the water again, and did it a couple of more times. I was not sure we could stay airborne because the nose did not go up and we seemed unable to rise more than a few meters above the water. Yellowknife Bay is fifteen kilometers long so there was no lack of space for a landing if we had to return to the dock and unload some equipment.

After a run of more than a kilometer, the plane became airborne, but when it was five or six kilometers from the dock it was barely fifteen meters above the water. At this point, Shaver eased off to 2,050 rpm, at half-flaps, in order to gain a little speed and more lift. Our air speed gradually increased until it finally exceeded 150 kph. Only then did we begin to climb. Instead of making a quick turn toward our destination, Shaver continued to fly into the wind and imperceptibly turned the ninety degrees toward Shoestring Lake.

Setting up base camp was a two day job with all personnel helping. We had eleven people who would live at base camp or operate from it. The personnel consisted of Keith Daybell (the party chief), myself, a cook, and two graduate geologists and two student assistants who would be ground prospecting teams. There were also four members of the air crew: the helicopter pilot and his mechanic, and the Beaver pilot and his mechanic.

The Shoestring Lake camp was located so we could explore an

area bounded by the Dismal Lakes to the north, Contwoyto Lake to the east, and Great Bear Lake to the west. All were within 250 kilometers of base camp, or within the effective operating range of the helicopter. The Shoestring Lake camp was just inside the treeline so that the forested area to the south was excluded from aerial prospecting.

In addition to its central location, Shoestring Lake had two other prerequisites for a campsite: a portion of the lake was ice-free, and there was a sandy beach adjacent to open water. Although there were several ice-free lakes up to thirty kilometers further north, they were shallow and had rocky shorelines. A sand beach is necessary for base camp because the plane must be securely grounded to be unloaded and loaded. Rocks of any size are extremely dangerous because they would puncture a float that is deeply pressed into the water when loaded planes takeoff and land.

Open water was also required because all of our supplies arrived in float planes. Planes equipped with a wheel-ski combination could have supplied camp by landing on the ice, which covered 90 percent of the lake. In mid-June the ice was still strong enough to support a loaded plane but, in three weeks, when base camp would move north, the ice would be too weak and the move might have to be delayed until breakup or until the plane was equipped with floats. Furthermore, if the camp's supplies were unloaded on the ice there was the problem of transferring them to the shore across a constantly widening moat of water. There was not enough personnel at base camp to do this.

There were two ponds of open water on Shoestring Lake, even though the average ice thickness was about a meter. The largest area of open water was at the southern end where the water was shallow. The sheet of ice covering the shallow water acted like a pane of glass in a greenhouse and converted sunlight to heat. The heat was absorbed by the dark boulders on the bottom and not dissipated in deeper waters. Ice quickly melted. This pond was not suitable for a camp site because there was no sand beach and because many large boulders projected to within a few centimeters of the lake's surface. Landings and takeoffs would be too dangerous.

Near the north end of the lake there was a second pond of open water. It was formed by a small river pouring a large volume of runoff water into the lake. At the latitude of Shoestring Lake runoff begins in late-May when daylight rapidly lengthens. The longer hours of sunlight warm the rocks of the surrounding hills and, except in ravines filled with drifted snow, the thin snow cover melts quickly. Runoff water collects sufficient heat from the warmed rocks to melt the ice where it enters the lake. The pond of open water at the river's mouth was about fifty meters wide and less than a kilometer long, but it was long enough for a fully loaded Beaver to land and long enough for an empty one to takeoff. Each day the opening would grow longer. Our only worry was a strong south wind. If it occurred, it could blow the sheet of lake ice against the sand beach and cover the open water. Fortunately, this did not happen.

The river carrying runoff water entered the lake in a small bay that had a sand bottom and a wide sand beach. The ground behind the beach was level and covered by a grove of five dozen widely spaced spruce trees. The ground between the trees was open, like a park, and covered by a thick carpet of caribou moss; and because the beach faced south, down the length of the lake, it caught all of the breezes that funneled up the lake. Flies were blown away and the dry beach sands did not breed mosquitoes. It was an ideal campsite.

Although Shoestring Lake is fifteen kilometers long and up to a kilometer wide, it was unnamed before we arrived. Its shoreline is a continuous escarpment of granite, except at both ends. Shoestring Lake occupies a major fault zone that glacial gouging had excavated into a lake basin. In heavily glaciated terrain on the Canadian Shield, major fault zones can be traced as straight or slightly curving lines that usually contain a chain of lakes. The Shoestring Lake fault zone is over 60 kilometers long and its surface expression is a series of pencil shaped lakes.

Shoestring Lake is a textbook example of a glacially excavated lake because the northwest-southeast direction of the fault coincided with the northwestern movement of the ice. The basins of these lakes have similar configurations: the shallowest portion is near the source of the ice, and the deepest part is found in the direction the glacier was moving. In the case of Shoestring Lake, the deepest portion of the lake is its northwestern end. Glacially excavated lakes have spoon-shaped cross-sections. In the illustration below, the ice was moving from left to right.

The spruce trees growing behind the beach were seven to eight meters tall, and obviously the same age. Although winter conditions

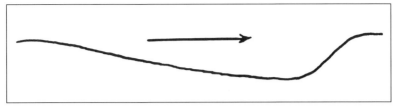
Arrow points in the direction of ice flow.

are still very harsh, the protected site allowed the trees to become relative giants compared to those that grow in clefts in the rock escarpments bordering the lake or on the higher, undulating ground behind the escarpments. The trees growing on the beach sands have the further advantage of growing on ground that thaws early in the summer. This means they have access to more mineral nutrients during a longer growing season than trees growing on the uplands, where ground does not thaw until later in the summer. Because the trees at base camp are widely spaced, they have branches down to their roots. They are giant Christmas trees that made very useful aerial masts.

In order to get the best reception and transmission, our assigned radio frequency required that our aerial be thirty meters long and be oriented north-south. We found two trees spaced the right distance and cut down two intervening ones. The stumps of these trees gave me the opportunity to examine their growth patterns. I examined a tree that was thirty-eight centimeters in diameter at a point fifteen centimeters above the ground. It had over 250 annular rings that were in three distinct groups: an inner core about three centimeters in diameter; a middle belt twenty centimeters wide containing 158 fine rings; and an outer ring fifteen centimeters wide that had forty-two rings. The rings in the heartwood were so close together that I could not distinguish them, even with the aid of a magnifying glass. I estimated it had sixty rings because, later, I cut down a sapling about three centimeters in diameter (and a little over one meter tall) that grew in an isolated grove in the barren lands. By whittling at a long angle across its stem, I was able to count fifty-five rings.

My observation of the very slow growth of dwarf trees in the barrens made me curious to compare their growth pattern with one of the giant trees we had cut down when stringing the aerial. From this initial observation and several related comparisons, I interpreted that there have been recent climatic changes in the western Arctic.

I have crossed the treeline many times, in helicopters, planes, and walking. It is indefinite. The treeline is drawn where groves of trees form more than 20 percent of the cover. The tallest trees in treeline groves are two to three meters high. But, most trees are dwarfs growing on the south facing slopes of hills or in protected valleys. At the extremity of the range of trees, only black spruce can grow, and they are knee-high dwarfs, in small groves, on extremely well-drained ground, at the bottom of hollows where snow drifts protect them from the harshest conditions of winters.

Trees over one meter high at the treeline have interrupted crowns. These trees have normal growth for the first meter and then a gap of thirty or forty centimeters, consisting of a bare trunk with a few stubs that had been branches. The top is a normal bushy crown. The branchless gap is caused by the abrasive action of snow crystals blowing along the snow's surface after drifts have filled the valleys where they grow. At sub-zero temperatures, snow crystals have sharp, cutting edges and, when propelled by 100 kilometer winds, they denude the needles from limbs just above the snow's surface, and the limbs die. Once a tree gets its crown above the drift line, an increase in height is comparatively easy, but few trees manage to do this when the winds buffeting them have an uninterrupted sweep of hundreds of kilometers.

According to the map, Shoestring Lake camp was 60 kilometers west of the treeline, which in this region runs north-south. The terrain east of Shoestring Lake gradually rises from elevation of 200 meters at base camp to 270 meters at Point Lake, 60 kilometers to the east. As the land rises, trees begin to thin and soon disappear except for widely scattered groves of dwarfs that precariously survive in micro-environments. For example, 120 kilometers east of base camp, we found a small grove of widely-spaced dwarf spruce trees growing on a sand flat where a small stream joined a larger one. This grove was in the lea of a high hill that afforded it protection from the wind. This was an optimum growing site.

The dwarf trees in this grove were growing on well-drained sand, similar to the sand at base camp. Winter winds had blown spruce seeds many kilometers across the snow crust and they were embedded in a drift in the lea of the hill. When the drift melted the

Basecamp at Shoestring Lake.

seeds lodged in the sand and germinated. They live a precarious life. The key factors to extending the range of trees into the barren lands are: growing sites on the best drained soil (usually sand), and enough summer heat to germinate seeds. Once trees have taken root, drifted snow is their best friend (provided it is not too deep) because it forms an insulating blanket that moderates the harshest effects of winter. When spring comes, runoff water percolates downward and carries enough heat into the sand to thaw the upper few centimeters. A porous soil, like sand or gravel, minimizes the effects of permanently frozen ground (permafrost) by allowing tree roots to begin extracting mineral nutrients from the soil before the snow cover melts.

While the trees are still buried in the snow they absorb heat from the sun. The snow melts several millimeters around the tree's needles. The enclosing snow is a blanket that retains a thin layer of air that is saturated with water vapor and carbon dioxide. The abundance of carbon dioxide comes from the bio-degradation of organic matter, lying on the ground or trapped in snow drifts, that has defrosted but is not yet exposed to the air. As the drifted snow melts, the tree's crown is exposed to direct sunlight. These factors create optimum growing conditions in an extraordinarily harsh environment so that dwarf spruce trees accomplish much of their annual growth while still partially buried in the snow.

The giant trees growing at base camp began life as dwarf trees growing far out in the barren lands. They remained dwarfs for sixty years but increased their rate of growth and their height when the climate warmed. During their early years, the trees growing on the beach at Shoestring Lake lived at the furthermost fringe of the treeline. The closeness of the heartwood rings was not due to the trees' living close together, as in temperate forests, where the struggle for life is competition among individual trees. The sixty rings in the heartwood represent the struggle to maintain life in an extremely harsh environment. The trees growing in the grove at the Shoestring Lake camp are as widely-spaced as the dwarf trees growing on sand flats fifty or sixty kilometers beyond the treeline.

Wide spacing is necessary to obtain maximum sunlight during the short growing season and is part of the most favorable ecological conditions necessary for dwarf trees to survive in the barren lands. At the fringe of the barren lands, all of a trees' energy is required for survival. Almost everything else is against their survival.

The wide spacing of the trees on the beach at Shoestring Lake is a growth pattern left over from the harsh climatic at the time they germinated. This same harsh climate is still present to the east, beyond the treeline and at higher elevations, but is no longer present at Shoestring Lake. In the past 200 years there have been two warming events in the western Arctic. They are recorded in two belts of tree rings around the cores of the trunks of the giant trees growing on the sand flat where base camp was located.

The middle and outer belts surrounding the cores show sudden spurts of growth. I counted 158 of them in the middle belt and 42 in the outer belt. These rings are evenly spaced and narrow but have distinct summer and autumn period of growth. The spurts of growth are called release. In general terms, release occurs whenever a tree can accelerate growth. During a tree's lifetime there can be more than one release.

In temperate forests, release usually occurs when a tree gets its crown into the upper canopy where it receives direct sunlight. Prior to this, a tree's crown is beneath the canopy formed by mature trees. Growth is slow and rings are narrow until a mature tree dies by disease or is blown over in a windstorm. The young tree is released from the shade of a mature tree and grows at a steady rate until it is weakened by disease, damaged by storms, or reaches senescence. But, in the harsh arctic climate, a tree's major competition is the weather, not other trees.

Release can only come from a general warming of the climate that allows dwarf trees to push their crowns upward so they can make more efficient use of light. To confirm this hypothesis, I made a second cut in the trunk of one of the giant trees we had cut down at Shoestring Lake. The cut was one and a half meters above its roots. I found 158 growth rings in the middle belt. These rings are the tree's first release. They indicate that, before its first release, the giant tree was a dwarf about one meter tall, resembling the dwarf trees growing far out in the barren lands.

The outermost belt had forty-two growth rings. This was a second release. Each ring in this belt was slightly thicker than the 158 rings of the first release. This indicates that the western Arctic became slightly warmer forty-two years ago, and the trees growing on the beach sands at Shoestring Lake again increased their rate of growth. I did two things to confirm this. I cut the top one and a half meters off of one of the giant trees we had felled, and found that this portion had grown in the past thirty-four years. To confirm that the second release was due to a further warming of the climate, I walked a kilometer down the shore of Shoestring Lake and felled one of the stunted trees growing in a cleft in the rock escarpment rimming the lake. This growing site was second only to the beach site for protection and good drainage.

The tree I cut down was about seven centimeters in diameter, and about three meters tall. It had a growth pattern similar to the two inner belts of the giant tree at base camp. It had an inner core of heartwood, slightly over three centimeters in diameter. The heartwood had more than sixty rings that were nearly obliterated because the wood cells were filled with dark brown resin. The outer belt contained forty-two narrow, but distinct, rings that could be counted. These forty-two rings corresponded to the outer forty-two rings of the giant trees growing on the beach. The outer belt accounted for about two-thirds of the tree's diameter and nearly three-quarters of its height.

The tree had germinated and grown at the same time as the giant trees on the beach were experiencing their first release. At that time the cleft became a site that would support tree growth. At the cleft site, it took more than sixty years to grow three centimeters in diameter and a meter tall. At the second release of the giant trees (forty-two years ago), the cleft tree underwent release and grew rapidly in diameter and height. I confirmed this by making a second cut in the trunk of the cleft tree one meter above its roots. It had a few fine rings in the center, but the outer forty-two rings made up most of the trunk's diameter.

With each slight warming of the western Arctic, trees extend their range into the barrens. The cycle of advance begins with a grove of knee-high, widely spaced dwarf trees just managing to exist at optimum growing sites beyond the treeline. With a slight climatic moderation, dwarf trees develop into normal, but stunted, trees, like those growing in the clefts of the escarpment rimming Shoestring Lake. And, on further warming, these stunted trees grow into normal trees like the grove at base camp. This interpretation is based on the ring patterns seen in cuts made at several places in their trunks and by comparing the spatial distribution of trees near the treeline and at Shoestring Lake.

If all the trees growing in the grove on base camp beach were cut down or destroyed by fire, the second generation of trees would grow closer together and there would be two or more species. After the climate moderated, closely spaced trees could survive on the same area of land, because the new competition would be among trees rather than with the weather. After black spruce, aspen is the tree species that is most tolerant of arctic climatic conditions. In multi-species forests, the competition for survival is each tree against all others, rather than a small number of dwarf trees of one species against the climate. This interpretation is further confirmed by observing the mixed groves of closely spaced spruce and aspen trees growing along the shores of lakes to the south of Shoestring Lake, where the arctic climate is in transition to sub-Arctic.

CHAPTER 2

Western Holiday

We completed setting up Shoestring Lake base camp on June 17th, and I made chopper traverses the next two days until Keith got the camp in running order. During these two days, the Beaver set out the two junior geologists in fly-camps where they began ground prospecting. Keith's immediate concern was seeing that base camp had the proper inventory of aviation gas and supplies, and laying out a systematic pattern of helicopter traverses. It was absolutely necessary to have a flight plan for all prospecting traverses made by the helicopter or Beaver. There could be no substantial deviations from the flight plan because, in case of mechanical trouble or bad weather, Keith had to know where to send a search and rescue party.

After base camp was in routine operation, Keith and I would share the job of helicopter and Beaver prospecting. Alternatively, Keith would use the helicopter, and the Beaver would take me to the shore of a lake in the morning and pick me up a kilometer down the shore in the late afternoon. I would spend the day making a six or seven kilometer traverse inland and another on the return. Keith would do all the helicopter prospecting until fatigue caught up with him. Then, I would relieve him for one or two days.

Just as we were about to begin routine exploration traverses, Keith was informed on, the evening radio schedule, that he would have to fly to Dawson, Yukon Territory the following day. A nickel discovery had been made and a staking rush was developing. One graduate geologist was to remain at Shoestring Lake and use the chopper on traverses already laid out, and the other, with his assistant, would come with us. Pete Dankas was selected to remain.

INCO's chief geologist wanted a quick report on its potential in order to see if the showing was important enough to rush another crew into the area. If our examination showed it be a significant discovery, the party chief was authorized to negotiate an immediate option on the property in order to keep it from being controlled by a competing company. We were the only crew that could get there on short notice and we would be there for an indefinite period.

Until 1955, Dawson was a dying town of 400 left over from the

Klondike gold rush of 1898. Since then, it has doubled in population because it is a jump-off base for oil and gas exploration crews working on the Peel Plateau, Eagle Plains, and Mackenzie River delta. We were not the first geologists to reach town in the wake of the nickel discovery, but we were one of the early arrivals and were best equipped to get to the showing and evaluate it.

Going to Dawson required a shift of 1,150 kilometers to the west, with a refueling stop halfway, at Norman Wells. The showing was eighty kilometers north of town, at Seela Pass, near the headwaters of the Blackstone River. It is mountainous country, with peaks as high as 1,375 meters and the major valleys at about 850 meters. The whole region lies along the eastern edge of the coast range mountains. Only the highest peaks, plus three of the major valleys radiating from the mountains, were glaciated. The glaciated valleys extended eastward from the eastern edge of the mountains like the arms of a squid. In mountainous terrain that has been glaciated, the major rivers flow in wide valleys with U-shaped profiles, which frequently contain a chain of lakes. The smaller side valleys and the middle slopes of the uplands were not glaciated. They have retained their original V-profiles, and have become hanging valleys (often with waterfalls) where they join U-shaped glacial valleys.

Unglaciated terrain has a very different appearance from glaciated terrain. The most striking feature of unglaciated terrain is the lack of lakes and dendritic, or trellis, drainage patterns. All of the major streams flow in V-shaped valleys. By comparison, glaciated terrain has a subdued topography (like the Canadian Shield) and is dotted with many lakes of all sizes and shapes, and the drainage network is highly disrupted so that one cannot always find the boundaries of watersheds.

The lack of lakes in the area of the nickel prospect was a serious problem since we had to find a place to make camp. The lightly glaciated uplands had some small lakes, but steep rock walls enclosed them. The only place to land was on the Blackstone River where there was a lake in the U-shaped valley. It was a scary feeling flying down the valley with cliff walls less than 300 meters on either side, but the main danger was from turbulent air rather than rock walls. The lake we landed on was eight kilometers from the prospect and 650 meters lower in elevation.

We landed at 9:30 in the evening and took three hours to eat and set up camp. Then, in the deep shadows of midnight twilight, we collapsed into our sleeping bags. We did not get started until ten o'clock the next morning. Keith estimated it would take him four hours of hard walking to get to the showing, while the other geologist and myself would take six or seven hours to make the trip. We would walk along the crests of ridges and see what we could see before arriving. If either of us saw anything of interest, we would collect a grab sample for assay and remember the location. If a second visit indicated that it had potential value, we would stake it using licenses we had purchased in Dawson.

Aerial photographs of the area were not available, but we knew the general direction to the prospect because Shaver had circled it three times before landing. Fortunately, it was above the treeline so that I would be able to see the showing from at least three kilometers away from the top of a parallel ridge. Furthermore, the claim owners had set up two bright orange tents on the showing, and had blasted four trenches in the gossan. I would arrive sometime in the afternoon so that the tents and the red oxide of the gossans would be highly visible.

It took me a little over seven hours to walk to the prospect, while the other geologist arrived an hour earlier. I saw and sampled three poorly exposed gossans on the way. When I arrived, I examined the mineralization exposed in the trenches. The unweathered mineralization consisted of about 10 percent sulfides, almost all of it pyrrhotite (an iron sulfide), with a few grains of nickel and copper sulfides. The main mineral in the gossan was chromite, an oxide of chromium. We were looking at a typical podiform chromite deposit that contained accessory sulfides.

Podiform chromite deposits are also called alpine deposits because they occur in mountain ranges or where mountain ranges have been eroded to their roots. They are usually concentrated along the edges of continents (within 250 kilometers of the continental shelf), and they are thought to be a feature associated with subduction trenches of oceanic plates.

Global mapping, since 1960, coordinated with global geophysical investigations, have defined fifty or more plates that form the earth's crust. They resemble the segments that form the surface of a soccer ball. Continental plates are rigid and average about 140 kilometers thick, but oceanic plates average forty to sixty kilometers thick and are flexible. Both continental and oceanic plates are continually in motion, moving at rates from one to fifteen centimeters per year. Where oceanic plates collide with continental plates, the oceanic plate is subducted (thrust) beneath continents. The tremendous force generated by colliding plates pushes up mountains over six kilometers high, and can compress the widths of continents by 150 kilometers or more. During collusion events, lasting millions of years, the rocks along the continental margins are crumpled like wet toilet paper and then pushed over one another like many shingles. This is the structure of the Canadian Rockies, from Calgary west to the Pacific Ocean and north to Alaska.

Oceanic rocks that are subducted beneath continental plates begin to melt at a depth of about 150 kilometers. The molten rock (magma) finds outlets as chains of volcanos inland from the compression front or as a chain of offshore volcanic islands (island arcs), like Japan and Indonesia. The magma comes to the surface as lava. Some of the magma, however, is trapped in deep chambers that lack outlets to the surface. This rock partially crystallizes into a mush. The heaviest crystals sink to the bottom of the chamber where they form layers. Chromite and sulfide minerals are very heavy and form bottom layers. The lighter silicate minerals, as they crystallize, concentrate at the top where they also form layers.

When the compressive forces of mountain building resume, these chambers are emptied like toothpaste squeezed from a tube, and the crystal mush within them is extruded toward the surface. The mush of crystals is usually injected into fault zones. Sometimes it forms large, irregular masses, but more often it is emplaced as a series of elongate pods that may or may not be joined. As this crystal mush is extruded toward the surface, the layers of chromite-sulfide minerals are shattered and jumbled. When this rock finally solidifies, before reaching the surface, the fragmented layers of chromite and sulfides are like raisins in bread.

What we were seeing was a podiform chromite deposit with accessory sulfides. Usually podiform chromite deposits are not large, but occasionally they are a source of chrome ore if they are located near transportation. Their nickel and copper contents are uniformly low and they are never large enough to be mined for their nickel-copper content. After the half an hour we needed to catch our breath, Keith distributed the assay samples he had collected into to our backpacks and we started for camp.

Keith estimated the grade of the richest portion of the pod at 0.4 percent nickel and less than 0.1 percent copper. However, we had no way of estimating its precious metal content (primarily platinum). We did not think it would be much and we had a very poor opinion of the economic potential of the mineralization we had seen, or any others that might be found in the area. Subsequent assays confirmed Keith's visual estimates.

On the morning of the 23rd, we did some prospecting near camp because there was a large area of exposed rock that looked like it might be of interest. It was not. We flew back to Dawson late in the afternoon so we could take a shower, have a few beers, and sleep on hotel beds. We were dead tired, so Keith declared the 24th a holiday. We slept until noon and after lunch visited one of the three placer gold mines operating in the district.

The Yukon Consolidated Gold Mining Company's dredge was the most accessible. It was on the Klondike River and a taxi took us to the foot of its gangplank. As geologists, the captain welcomed us aboard and explained its operation. The dredge was a rectangular steel barge twenty-five meters long and twelve meters wide. It floated on a pond of its own making that was barely larger than the dredge. At the front end was a continuous chain of digging buckets mounted on a massive steel boom fifteen meters long. The boom was buried in the water and the moving chain of buckets scooped up the gold bearing gravel and dumped them aboard the dredge. The coarse pebbles and gravels were screened and dumped on the spoils pile at the rear.

The remaining particles were classified into grit, coarse sand, and fine sand. These uniform sized particles were fed across the surfaces of vibrating trays one meter deep that were filled with steel balls. A strong upward current of water rose through the balls. The light particles floated on the upward surge of water and flowed across the top of the tray and over the edge. Heavy minerals sank against the rising water, became entrapped among the balls, and gradually worked themselves downward to form a concentrate. The first tray in the series was filled with marble-sized ball bearings, and was designed to trap gold nuggets and grit-sized heavy mineral particles. The last tray was filled with BB-sized pellets that trapped gold dust, as well as particles of other heavy minerals.

The heavy mineral concentrate is a reddish-black sand consisting of fragmented crystals of garnets, magnetite, and chromite. Gold forms a very small fraction and is invisible unless there is a large nugget. The concentrate is trucked to Dawson and the gold recovered there. The dredge recovered less than half a gram of gold (worth about sixty cents at 1960 prices) from every cubic meter of gravel and, because of the severity of the winter, it operated only about 150 days per year.

Gold bearing gravel results from the weathering of mineralized rocks that are too low grade to be mined. Gold is freed from the enclosing rocks by mechanical weathering and, as it is transported downstream, water agitated gravel shapes gold particles into rounded nuggets, flakes, and dust. They, and other heavy minerals, concentrate in the sand and gravel at the bottom of rivers and streams. When an area is glaciated the residual gravels are dispersed, but Dawson was not glaciated and the gold bearing gravels remained in place. They were, however, covered by glacial-fluvial sediments that originated from the continental glacier whose western edge was thirty kilometers to the east.

During the tens of millennia in which the glacier existed, spring runoff water flooded the Klondike valley and deposited variable thicknesses of unconsolidated sand and gravel on top of the gold bearing gravels. In time, the valley was nearly filled to the top and the severity of the winters permanently froze the water trapped in the pores of the gold bearing gravels. When the last glacier melted, the vast volume of meltwater cut through the covering layer of glacial-fluvial gravel and partially re-exposed the gold bearing gravels. Sourdough prospectors found gold in 1897.

Permanently frozen ground (permafrost) occurs in non-glaciated as well as glaciated areas, and is often found thousands of kilometers from the nearest glacier. It forms whenever surface temperatures are below freezing for a large part of the year, and especially where there are strong winter winds and cool summers. In parts of Western Canada and Alaska, these conditions are prevalent and permafrost is still forming. In other places, it is slowly disappearing.

Permafrost forms most easily in organic deposits that are poorly drained. Plant life in the harshest arctic climate is mostly ground hugging sedges, mosses and lichens that grow on a soil that thaws to a depth of less than one meter during the summer. When these plants die, they only partially decay because of the shortness and coolness of the summer. During succeeding summers layers of half-decayed, spongy vegetable matter accumulate. The water in it freezes and the thickness of permafrosted ground builds up.

After we had inspected the dredge, we walked up the valley to see the monitor in operation. A monitor is a giant water nozzle mounted on a heavy tracked platform whose job is to expose the gold bearing gravel for the dredge to process. Only gravel that occupies the center of the Klondike valley contains gold and, in order to expose it, the overlying ten meters of sediments must be removed. This is efficiently done by the monitor. It shoots a jet of water against the frozen ground and, when the water hits loosely frozen gravel and pebbles, they explode in all directions.

This is not true of the peat and silt lenses that had retained water. The ice in them acts as cement and they must be undermined before they disintegrate. All the loosened material flows into an artificial channel that dumps it into the river behind the dredge. The monitor operates only during the spring and early summer when there is an abundance of runoff water. The water it uses is stored behind a dam several kilometers upstream and fifty meters higher in elevation. It arrives by pipe. The fifty meters of head gives the monitor's water jet enormous cutting power that can move huge amounts of material in a short time.

After the top of the gold bearing gravel is exposed, bulldozers build a three meter high dam across the stream to impound the reduced summer flow of the Klondike River. During the winter a sheet of ice covers the lake and insulates the water from losing the heat it accumulated during the summer. The heat stored in the lake water gradually melts the permafrost in the gold-bearing gravel. It requires three years of soaking beneath the water of the artificial lake for the permafrost to disappear from the gold-bearing gravel. The gold-bearing gravel averages about seven meters thick, but this thickness is highly variable. It can be as little as two meters thick over ledges to as much as twelve meters in potholes. The dredge captain carefully searches for potholes, because they often contain bonanza accumulations of gold dust and nuggets.

At the outer edges of the auriferous gravel, the overlying sediments that had not been disintegrated by the monitor stood as a permafrost wall. We had the opportunity to see the wall in cross-section. About three-quarters of the capping material was gravel. The rest was sand, poorly sorted silt, and frozen peat that formed irregular lenses. There were also thin lenses of ice. The color of the ice appeared to be black because thawed peat above it formed a black muck that dripped over it, but beneath the coating of peat the ice was clear and white. The ice originated during late autumn when water from a small tributary stream spread a thin film of water that froze. In the following spring, it was buried by sediments before it could melt. The peat was mostly the remains of moss that had partially decayed and, when re-exposed, it resumed decay and gave off a fetid odor.

I asked the monitor operator if he had ever uncovered mammoth bones from the frozen ground. He said that mammoth bones were quite common, or at least he thought they were mammoth bones because they were large. Several professors from the Universities of Toronto and British Columbia had visited him over the years and had told him the bones were from mammoths. He showed us a couple of large teeth and gave them to us as souvenirs. Then, he pointed to a place in the gravel wall where mammoth bones were protruding. He said he had never found a complete skeleton but that this one could possibly be whole. He had found one of the tusks that he showed us. It was about two and a half meters long, measured along its curve, and had begun to turn inward. The ivory was chipped, discolored, and cracked, and had no commercial value; and he added that he had never found one worth selling for ivory. He had, however, sold them to tourists.

He turned the monitor at a different angle so we would not get drenched by spray or hurt by flying rocks, and the party chief and I went in the direction he indicated. The other geologist was not interested because he did not want to get his feet soaked in the frigid, gritty sludge that the monitor had created. We inched along the permafrost wall trying to find firm footing but we soon slipped. Cold water filled our boots. We quit trying to keep dry and started looking. I found a leg bone and we looked in the immediate area for the rest of the skeleton, but the mud spattered surface concealed everything except bones that protruded a long way. I sloshed through the sludge to the monitor and got a bucket, and we began washing down the gravel wall near the bone. It took six buckets of water to expose other bones, but it was worth it.

We found a tibia, fibula and some metatarsals sticking our from the wall, all held together by cartilage. I found a toenail attached to one of the metatarsals, and a scrap of skin that was covered by coarse reddish-brown hair. Further down the wall, we found a set of vertebrae with a couple of ribs but they were strongly embedded in the frozen ground and could not be dug out. The bones we saw probably belonged to one animal. They were probably carried a short distance downstream from where the animal died and deposited in deep water at a bend in the stream.

I will try to reconstruct how the mammoth might have died and become embedded in sediments. It was probably a lone bull that was attacked by a pack of wolves, during late autumn, while it was feeding on scrub willows and sedges growing along the river. It did not have time to run to the crest of a hill, which would have been its best defense, so it went into the middle of the river. The river was wide and there were no banks from which the wolves could leap at its legs to hamstring it, but in autumn the river was a trickle. Water did not afford much hindrance to the slashing attack of the wolves.

After a protracted battle, the wolves managed to hamstring the mammoth. It fell back on its haunches. Then, they blinded it. Then they began to eat, probably starting on the hindquarter where the whip motion of the trunk could not reach them. After it was dead, the alpha dog and alpha bitch of the pack gorged themselves on the reserve of fat stored in the cap-like bulge on top of the mammoth's skull. Thereafter, for more than a week, if not driven off by lions, the pack feasted. The mammoth was reduced to a pile of partially articulated bones and a few scraps of skin. Soon after the first snow, the river ceased flowing and the bones were frozen in place. During spring runoff, the bones were freed from the ice and transported to the nearest deep hole where they sank in a jumbled heap and were covered by sand and gravel. During the winter, they were embedded in permafrost.

The mammoth (Mammuthus primigenius) became extinct about 10,000 years ago. We know this from radiocarbon dating. All living organisms continually replenish dead cells and, in the process of replacement, they maintain a constant ratio between the three carbon isotopes: carbon-12, 13, and 14. Carbon-12 accounts for 99 percent of the carbon in living cells, carbon-13 and 14 for about 1 percent. Carbon-14 is continuously created in the uppermost parts of the atmosphere by cosmic rays bombarding nitrogen-14 and transmuting it into carbon-14. Carbon-14 is radioactive and has a half-life of 5,700 years.

When an organism dies, the ratio between carbon-12 and carbon-14 is no longer maintained because carbon-14 is no longer ingested into the body from the food it eats. Carbon-14 in the mammoth's bones and flesh decays at a constant rate, so that half is gone at the end of 5,700 years, and by 40,000 years not enough remains to be detected. By measuring the amount of carbon-14 still in flesh or a bone, the time of the organism's death can be accurately determined, provided it died within the past 40,000 years. The latest radiocarbon date for a mammoth bone is about 10,000 years, about 2,000 years after man entered North America by migrating from Asia across the Bering plain—now the Bering Sea.

The mammoth's nearest living relative is the Indian elephant, although they do not greatly resemble one another. Mature mammoths stood as high as their living African cousin (three meters), but they lived in an environment similar to that of the caribou. In the winter, mammoths ate the scrub trees and bushes of the sub-Arctic (taiga), and in the summer they grazed in the barren lands. Judging by the quantity of bones that have been preserved in Siberia, mammoths were the most numerous of the large mammals (mega-fauna) in the Arctic and sub-Arctic during the glacial eras.

There are several islands in the Arctic Ocean, near the mouths of the Kolyma and Lena Rivers in Siberia, that are wholly composed of sand, ice, plant remains, and mammoth bones; and, occasionally, complete carcasses have been found that were so well-preserved in permafrost that their meat was fed to sled dogs. Mammoth ivory from these locations was a common item of trade (measured in tons) during the nineteenth century.

Mammoth bones have also been found in England, Central Europe, and from coast to coast in the United States. At several sites, spear points have been found among the bones of mature mammoths. Man probably exterminated them, as man probably exterminated all the other giant indigenous mammals in North and South America. These mammals disappeared in an instant of geologic time, after the end of the last ice age when the first humans migrated to the Americas from Asia. Cooperative hunting tactics provided abundant feasts for the earliest hunters, after they drove

Mammoth.

herds over cliffs, or into lakes where they were speared from canoes. After a herd was killed, the hunters gorged themselves on meat for a week or more, or until the flesh rotted, and then went looking for more.

From 1910 to 1970, the barren land caribou herd declined from 10 million to a low of about 270,000 in 1955. Currently, the Canadian herds are estimated to contain about 700,000 animals. North America's Aboriginal people are the principal cause of the herd's decline, because they slaughtered them in huge numbers to feed their sled dogs. Native fur trappers depended on sled transportation to run their trap lines.

The northernmost Indian bands live at least 100 kilometers south of the treeline. At the end of the summer, fur trappers often go to the treeline by canoeing on a large lake system or a major river. Dogs and a sled are part of their baggage. When they get to the treeline, or near it, they build a cabin and wait for the annual migration of the caribou herd. Most dog food comes from caribou that are mass killed during the autumn migration, from the barren lands to their winter feeding grounds in the scrub forest south of the treeline. Caribou can be slain in large numbers at one location where herds funnel between two large lakes in order to cross a river. This is an ideal location for hunting.

Natives spend the winter trapping arctic fox and wolves in the barren lands and mink, otter, fisher, and sable in the taiga to the south. Winter is the best time to trap fur bearing animals because their fur is in prime condition, and because sled travel to lay out trap lines is relatively easy on the wind packed crust of snow on ice covered lakes. Indian trappers always bring their families. Wives are needed to dress skins that are brought back to the cabin as frozen carcasses. They chop a hole in the ice and put the carcasses in the water until it thaws. Then, they skin it and stretch the pelt on a frame to dry.

Two or three dog teams, plus pups and extras, consume a prodigious amount of meat during the winter. Each dog eats about half as much meat per day as the average person. In early spring, additional caribou are killed during their northern migration. At one time, among Native trappers, the number of dogs that could be fed was a sign of prestige and affluence. As long as the price of fur was high, the caribou slaughter continued. Only a drop in the price of fur prevented the barren land caribou from facing extinction, like the mammoth before it.

The next day, June 25th, we returned to Shoestring Lake and resumed our planned prospecting traverses.

CHAPTER 3

A New Arctic Hand

While we were in Dawson, Pete discovered an extensive zone of gossans about 120 kilometers east of camp, well out in the barrens. He collected assay samples, but found no visible economic mineralization. However, the area was large and he saw many small rusty patches that he did not have time to sample. Two days after we returned, the graduate geologist and his student partner, who had been with us in Dawson, were set out to prospect and sample the entire zone.

Pete was given a day off from the fatigue of flying. Keith took the chopper on a traverse, and I used the plane to fly north to the Tree River country to see if there was any open water. We had been told that the mouth of the Tree River usually had open water early in the spring because tidal action broke the ice on the estuary and the runoff flood melted it. This information came from an outfitter in Yellowknife who flew fishing parties there to catch record sized arctic char. If any open water was present, our next camp would be there.

About half of the flight was a useful exploration traverse, but prospecting ended 150 kilometers north of camp because snow still covered everything but hilltops and the southern exposures of valleys. However, I found that the estuary of the Tree River was open, although Port Epworth was still completely ice-covered. The Tree River estuary would be an excellent campsite because it had several kilometers of sand beaches and its channel was deep.

The next morning, bad weather shrouded Shoestring Lake and we were confined to camp. We had fog in the morning, followed by an intermittent drizzle, but by one in the afternoon there were indications that the weather would clear. By three in the afternoon, the sun was at full power and the bush dried rapidly. We had the opportunity to look for arrowheads in the beach sands. Natives always camped on sand beaches for the same reasons we did. Rocks were just as fatal to bark and skin canoes as they are to the aluminum floats of our plane. At beach camp sites the Indians made arrowheads whenever they had bad weather. We found several piles of flakes and about two dozen points, ranging from small ones, used on birds, to eight centimeter caribou heads. About half of them were broken.

The day off also gave us the opportunity to take some pictures of camp. Pete and I went up the escarpment to get a panoramic view and, after we had snapped a few pictures, I looked for a patch of open ground. I wanted to show Pete the unstable nature of what passes for soil in the Arctic, and what happens when this material is disturbed by having a vibrating weight placed upon it.

Soil is a very complex material. It is formed by the mechanical and chemical weathering of the surface layer of rocks or from material derived from this rock. It is made up of an intimate mixture of clay, sand, free water, organic matter, air, and chemical compounds in solution, all in variable quantities at different times of the year. Clay minerals are an essential ingredient of soils. Clays have a high percentage of chemically combined water, and because they are ultra-small particles they have very high surface tensions. This causes them to stick together, giving clay a plastic consistency.

Most of the arctic portion of the Canadian Shield is covered by low hills composed of some variety of glacial till. Tillite is an unsorted mixture of boulders, pebbles, gravel, sands, and rock flour that was dumped on the land surface when the glacier melted. The rolling hills that result from dumping are called ground moraines. There are areas where bedrock predominates and equally large areas of unconsolidated sand. In some regions, there are extensive peat bogs and in other areas, boulder beds cover many square kilometers. Boulder beds are a surface layer of rocks up to house size. They may have moved only a few meters or a few kilometers from their place of origin, and they rest directly on parent material. There are almost no small-sized particles binding them.

When the last continental glacier was at its maximum size, it extended 180 kilometers south of the lower Great Lakes. When the southernmost portions of the glacier melted, it deposited tillites that are, in every respect, similar to the tillites at Shoestring Lake. Warm rains, falling on the tillite, oxidized the minerals containing lime, potash, and phosphorus. These mineral nutrients go into solution, and adhere to the surfaces of clay minerals or are absorbed by organic matter. This preserves them for plant growth. The availability of dissolved mineral nutrients in the soils in temperate climates is a principal reason why they support a diverse and luxuriant plant life. This cannot happen in the Arctic until permafrost thaws, rock flour hydrates into clay, and mineral nutrients go into solution.

One of the more curious aspects of the Arctic is the scarcity of soil in the scientific sense. At Shoestring Lake, only a thin surficial layer of soil had developed on the tillite, and this soil was present only where rock flour was the surface material of moraines. Rock flour is a characteristic product of glacial erosion. A glacier erodes by mechanical abrasion. No chemical weathering takes place in the frozen environment beneath a glacier because there is no free water. A glacier can pluck house-sized boulders from the tops of hills or grind the hardest rock into a powder finer than dust. After a glacier melts and tillites are dumped on the land surface, rock flour often collects in surface depressions where it is exposed to oxidation, hydration, and solution weathering. In the Arctic, however, weathering takes place at a very slow rate. Under arctic conditions, rock flour does not easily hydrate to form clay minerals, nor do most minerals readily go into solution. Summer heat thaws only the upper meter of tillites, and the ground below remains permanently frozen.

Permafrost is deepest beneath high hills that are blown free of snow and thus continually exposed to chilling winter winds. Permafrost effects bedrock, as well as tillite, by freezing its normal free water content of 2 or 3 percent. The depth of permafrost beneath the rock scarps rimming Shoestring Lake is probably 250 meters, while it is probably 150 meters beneath valleys filled with drifted snow. Patches of permafrost are found as far as 300 kilometers south of the treeline so that the presence of large trees is no guarantee that the land is free of permafrost.

Trees growing on glacial till underlaid by permafrost get their nourishment from the surface three to seven centimeters of ground. This is the thickness of much of the soil in the Arctic because this is only place where clay minerals can form by the hydration of rock flour. The feeding roots of arctic plants are concentrated in this layer. No arctic plant has a taproot because no mineral nutrients are available beneath the uppermost three to seven centimeters of soil, where rock flour has hydrated into a semi-plastic skin of soil over the rock flour beneath. The skin of clay minerals is a very distinct layer because it has a different color, texture, and strength from the underlying rock flour.

Underneath the surficial few centimeters of soil, only one meter of thawed ground is produced annually by summer heat. The rock flour particles in this layer are unaltered because they have not hydrated into clay minerals, and they cannot hydrate until they are exposed to higher temperatures. Rock flour has none of the cohesion of clay minerals. It is like a layer of marbles that are 1/100 of a millimeter in diameter. Thawed rock flour is like BBs packed in a barrel of water. Any downward pressure displaces it, and it swallows up any object placed on it. Alternatively, if it is shaken (by an earthquake, vibrations from a motor, or a man-made explosion), and the confining skin is broken, thawed rock flour liquefies and flows.

We once placed a diamond drill on a tillite wholly composed of rock flour. The vibrations of the diesel engine caused the whole rig to sink one meter into the ground, until it came to rest on permafrost. This occurred twenty minutes after drilling began, in spite of the drill resting on a platform supported by eight wooden sills. Usually, however, there is enough sand, pebbles, or boulders mixed with rock flour to keep it stable when it thaws, provided it is not disturbed.

One of the first things we show new men in the Arctic is the instability of tillite composed of rock flour. The novice is placed on a patch of tillite and other members of the party jump up and down like drunks. The novice looks on in amusement until the ground begins to heave and roll, as if it were made of jelly. A depression two meters in diameter forms and water begins oozing from the ground through small cracks in the clay skin and collects in a pool at his feet. Then larger concentric cracks appear around the edges of the depression and rock flour flows into the depression like lava. A slap of the hand on the clay skin causes an undulating roll.

There is no danger of being engulfed as long as the clay skin beneath the boots is not punctured. Two minutes after a person steps out of the depression, the ground rebounds and congeals and is perfectly stable to walk on. But, if boots break the crust and a person continues to jump, in less than a minute his boots will be in over his ankles. At this point, the quickest way for a person to extricate himself is to step out of his boots.

While we were on the hill taking pictures, I decided to show Pete the wonders of rock flour. I found a large patch of ground that was free of boulders and gravel and told him to start jumping up and down, as I was doing.

He thought I was crazy and said, "Man, don't play games with me!"

"Stay put and watch," I said, and continued to jump up and down.

He watched with a silly grin on his face, thinking that an old arctic hand was trying to make him play the fool. He was not going to be led on.

After about two minutes, a small area at my feet began to move and I began jumping in a widening circle. The ground began to visibly heave. Pete looked at the ground in amazement and then he stepped into the circle of heaving ground and began jumping up and down. The heaving motion fascinated him and he began to expand the area until he was in the center of a quaking circle two meters in diameter.

He gave me his camera and said, "Take a picture."

He was like a small boy with a new toy. He purposely allowed his boots to break through the crust up to his ankles so he could have a picture of his feet in the mire. Then he tried to pull his boots out, but suction held them. I told him to step out of them, but he did not want to get his feet wet in the water that had collected around his feet. He began to struggle to pull them free.

He fell backward into a sitting position, then forward, to try to loosen the suction. He failed. He sat down with a grimace of panic on his face. I took my knife from its sheath, walked over and cut the laces on his boots. Then I put my hands under his bent knees and pulled each foot from its boot. Then I collapsed his boots and pulled them from the mire. When we were back in camp we had a hot cup of tea with two tablespoons of honey in it. This is the best cure for chills after being caught in a cold rain or any other occasion when your body needs to be quickly warmed. The cook also warmed a pan of water and Pete bathed his feet in it. Then, he washed his boots to remove the mire. He was now a fully initiated arctic explorer.

CHAPTER 4

Breakup

The weather was clear, the sky was high, and there was little wind in the two weeks following Pete's learning experience. It was ideal flying weather and I remained at base camp using the chopper, while Keith was in Yellowknife chartering a plane to help move base camp to the Tree River and return with supplies. Warm weather had melted the snow cover on the hills south of Coronation Gulf, allowing us to move a week sooner than planned.

Three days before we broke camp, we witnessed breakup on Shoestring Lake. Breakup is the overnight disappearance of the ice sheet covering a lake. In order to understand how ice can disappear overnight, something must be told of the freezing and melting process of fresh water. Both freezing and melting of ice sheets on lakes takes place on the underside. Fresh water is densest, and therefore heaviest, at four degrees. If the surface water is cooler (as it is in the winter), or warmer (as it is in the summer), it forms a lighter layer on top of the four degree water. The water in arctic lakes is distinctly stratified.

During the winter, ice thickens downward because the thin layer of cooler but lighter water (between zero and four degrees) is directly beneath the ice. The cool layer beneath the ice does not mix with warmer water below because the ice sheet prevents wind generated convections. Wind blowing across the ice sheet continually removes heat from the transition layer. This causes two things to happen: 1) the layer of cool water directly beneath the ice increase in thickness; and 2) the ice sheet grows downward as the cool layer of water gives up its remaining heat (latent energy) and freezes.

The process reverses in the spring. The layer of water beneath the ice absorbs heat from sunlight passing through the ice (greenhouse effect), warms the underside of the ice, and the ice melts from the underside upward. Alternatively, warmer runoff water flows into the lake and the turbulence of its entry creates mini-convections that transfers heat to the underside of the ice. In both cases, melting occurs from the underside upward.

Melting ice on arctic lakes is a slow process because: 1) the ice is more than one meter thick; 2) most lakes do not have large influxes of runoff water; 3) the ice sheet prevents the wind from creating convection currents that bring warmer bottom waters (at a temperature of four degrees) into contact with the underside of the ice sheet. Without a convection, lake water remains stratified and the huge amount of heat stored in the four degree water is not available to melt the ice sheet; and 4) the surface of the ice sheet reflects a high proportion of sunlight, which means that only a small percentage of the sun's energy is available for absorption by the water beneath the ice sheet. Even when the sun is shining twenty-four hours a day, as it does in late spring and early summer, relatively little of its energy penetrates the ice.

The heat transferred to the undersurface of ice sheets is exchanged by mini-convection currents. At the beginning of the melting process, the ice's undersurface becomes dimpled. As melting proceeds, the dimples become pits and the pits gradually extend upward as tubes. The tubes are called candles. Candles grow upward into the ice and, at the same time, enlarge their diameters. Candles accelerate melting because vertical tubes of water are more efficient than ice sheets at intercepting sunlight and converting it to heat. By the time the thickness of the ice sheet has been reduced to less than a meter, it is honey combed with candles and has lost much of its strength.

While candling is going on, other changes are taking place in the horizontal dimensions of the ice. The first visible change is the ice sheet coming up. Ice comes up after the ice melts around the shore. A moat of open water along the shore indicates that the ice sheet is no longer frozen to the bottom. It floats up. This generally coincides with an influx of runoff water that can raise the lake level by half a meter or more. After the ice comes up, the puddles and slush on the ice sheet drain into the nearest cracks. The ice that is exposed is hard, blue, and has a dry surface.

Soon after the ice comes up, the solid sheet breaks into several large floes along existing cracks. The cracks were formed during the winter by gale force winds blowing across the ice and generating long, undulating waves in the ice. The ice fractured at critical stress points. These points occur where the vertical movement induced into the ice by undulations cannot be absorbed. The ice cracks. As long as the weather remains cold, the crack are sealed but with the coming of warmer weather water enters the fractures and lateral melting begins.

By the first week in July, the ice on Shoestring Lake had broken into a dozen very large floes with as much as ten meters of open water between them. The ice was rapidly melting along the edges of the floes, and slivers of candled ice were continually being broken off by rippling water. The slivers collected along the edge of the floes and, when they bumped into each other, they produced a tinkling sound like many small silver bells. If the floes are not disturbed by waves, they continue to melt until the ice is black and so weak that a person walking on it would drop through. At that stage, the ice would consist of vertical slivers of ice up to a meter long that are barely held together. This is the final stage of melting, but it is seldom reached. At the end of the first week of July, there was still plenty of strength left in the ice. Two days before, we had canoed out to it and batted softballs.

Sometime during the evening of the 11th, the wind shifted to the north and the hot weather abruptly ended. Fleecy clouds were replaced by a thin misty overcast. The wind's velocity kept increasing and, at one o'clock in the morning, the mechanics roused camp when gusts reached ninety kilometers per hour. Four 10 gallon drums full of gasoline were placed on each of the chopper's carrying racks, and the rotor blades were tied down. Then we pulled the plane high on the beach tail first, rolled two, full forty-five gallon gas drums under each wing, and tied each drum to the plane twice, first through the struts and then over the wings. In high winds, planes try to fly by themselves and end up being flipped on their backs.

The open water in front of base camp was whipped into meter high waves topped by whitecaps and blowing foam. The area of open water grew very rapidly. Large floes split into smaller floes, and the smaller floes were fragmented into cakes that were ground against one another until they disintegrated. This crunching action produced a low moaning roar over the whine of the wind.

When we got up for breakfast, there was no ice on the lake even though it had been 80 percent covered the previous day. As previously mentioned, most of the lake's water was at a temperature of four degrees centigrade, but the ice sheet prevented the wind from mixing warmer bottom water with the cooler water just beneath the ice sheet. After the wind fragmented the ice, it generated huge convection currents that brought warmer bottom waters to the surface. Ice fragments and small cakes were bathed in warmer water. The bottom water exchanged its surplus heat and melted the ice in a matter of hours.

The only ice that survived were a few floes that had been blown ashore at the south end of the lake. Some floes had slid over one another and were pushed as far as five meters beyond the shoreline. There they formed a haphazard wall nearly two meters high. A few small cakes were also perched on flat-topped boulders protruding from the water. This ice remained long after we left Shoestring Lake because the ice-air contact is a very slow and inefficient way of melting ice.

The visible phenomenon of rapidly melting ice was an indication of an invisible phenomenon that was just as dramatic. The convection that brought four degree bottom water to the surface also renewed its content of dissolved oxygen. The renewal of dissolved oxygen is necessary for fish living in the deep water to survive the summer. Only a strong wind can generate convections that can extend fifty meters or more into the water. The spring convection, which we witnessed, re-oxygenated the water. During the summer, after the ice sheet melts, twenty-four hours of daylight rapidly warm a surface layer of water. This layer may be as thick as ten meters and as much as twenty degrees warmer than the underlying water.

The reverse occurs in late autumn. The upper layer of water loses its heat. Shallow lakes lose heat very rapidly and they are the first to freeze. Heat loss in larger lakes is slower and they are the last to freeze. Just before freeze-up, when the surface and bottom waters are at temperature equilibrium, change of season gales power the autumn convection. The water is again oxygenated.

Spring and autumn are the only times during the year when lake water is not temperature stratified. Put another way, the only time deep water convections can take place is during the short periods at breakup in the spring and freezeup in the autumn. During these periods, lakewater has a uniform temperature and convections powered by gale force winds renew the oxygen content from top to bottom. Convections bring deep water to the surface, where the interface of atmosphere and spray blowing from the tops of waves resaturates the water with life sustaining oxygen necessary for fish to survive the long winter beneath the ice sheet.

CHAPTER 5

Scrounging

The first rule of the bush is to scrounge any useful equipment wherever you find it. Any abandoned camp is open for occupation, or if any of its equipment can be used elsewhere, it can be permanently borrowed. When I was on prospecting traverses, I always looked for abandoned exploration camps in hopes of acquiring some material from old buildings that could make our base camp a little more comfortable. Particularly, we examined all known mineral occurrences shown on our maps. There were not many.

A limited amount of prospecting had been done as far north as the treeline in the late 1930s, just before the beginning of WWII, and a small additional amount from 1946 to 1950. Gold was the only metal prospectors were interested in. In this isolated region, gold was the only metal that could be profitably mined because only gold could bear the high cost of air transportation to civilization. The earliest prospectors ignored all other mineralization, even if there were indications that it was present in bonanza quantities. INCO prospected some of the ground in the vicinity of the treeline, which had been examined by gold prospectors, in order to see if they had overlooked any base metal occurrences.

On July 14th, Keith decided to move base camp from Shoestring Lake to the mouth of the Tree River. The two prospecting teams were brought in so that all of our manpower was available for the arduous job of moving in one day. We began breaking camp at 6:00 a.m. All of us helped take down tents, lug them to the beach, box all of the cook's supplies, and pack our personal belongings. The first planeload left at seven o'clock, with a tent and the two men who would remain and erect it. I stayed at Shoestring until noon providing needed labor.

After helping to load the second plane and helping lug the plywood flooring of the third tent to the beach, Fred, the chopper pilot, and I began an exploration traverse to Tree River. I made the flight because Keith had to stay and make a gas and grocery inventory, and see that some gas remained at Shoestring for emergency refueling. Our traverse would take about six hours and we would arrive about the time the fifth plane load of equipment arrived. It would carry the last tent, and we would be relatively fresh and full of energy. It would be our job to erect it.

I planned a dog-leg in our traverse. I wanted to visit a gold showing on the north shore of Point Lake, near its western end, and spend half an hour making an intensive aerial survey in the area immediately surrounding the showing, as well as examining the showing itself. The prospect was located in flat country. When still ten kilometers away, we noticed strange protruding objects by the lakeshore. It was not until we had cut the distance in half that we saw that it was a camp consisting of five tent frames open to the sky, and one that still retained its canvas roof. We presumed someone was there, probably a prospecting party. Not until we were almost on top of it did we see that the tent was ripped and bleached white so that next winter's winds would shred it.

We circled the camp to confirm that it was vacant, and then spent thirty-five minutes making an intensive survey of the surrounding thirty square kilometers. We were searching for any small blotch of the slightest rusty zone that would indicate low grade sulfide mineralization of the type associated with gold. We found nothing so we flew back to the deserted camp to see what we could scrounge.

We landed on the nearest tundra hummock and appraised the situation. The tent frames were constructed of two-by-four lumber with waist high sidings made of fiberboard. Half of these had been blown from the frames and all of the flimsy wooden doors were unlatched and swinging in the wind. In one of the tent frames, the stringers that supported the canvas roof had collapsed and twisted the rest of the frame into a jumble of warped lumber. The floors were quarter inch plywood that sagged dangerously when walked on. After a few more seasons, they would be so weak that the next person walking on them would punch holes through them.

One tent had been used as a kitchen and mess hall, one was a warehouse, another was a sample preparation office, and three had been bunk tents. The crew that had sampled the gold showings probably consisted of twelve men. During the summer, they drilled half a dozen short holes, blasted a dozen trenches in slightly mineralized outcrops, and intensively prospected the area ten kilometers inland from the shore of Point Lake.

The first frame we entered was the geologist's sample preparation office. It contained two well-made, four-legged stools, and a small table of light, but sturdy, construction. We placed them outside the door so we could collect them when we returned to the chopper. Our tents at base camp lacked these luxuries and, since our traverse was headed toward base camp, we could carry some bulky items on the side racks. Fred coveted the table and one stool. He wanted them in his tent where he and the mechanic could use them to fill out flight reports and maintenance logs. No longer would he have to do his paperwork bent up on his cot with a board propped up on his legs half a meter from his face.

We also found a decayed cardboard box full of test tubes of the types used to make gold assays. We left them undisturbed. We had no use for them. I also found a steel ball, that was four centimeters in diameter, used to grind rock samples to prepare them for assay. I dropped it into my pocket. There were also some rock specimens on the shelves in which I took a professional interest but saw no visible gold. On the way out, we noted a faded paper sign tacked on the back of the door: Arctic Gold Mines Ltd.

I remembered the outfit. It had been active ten or twelve years earlier. It was a promotional company. The organizers had optioned this remote, minuscule showing and had raised approximately $500,000 from the public. The maximum amount they had spent on the property was $150,000. The rest was spent on themselves, at least that was my appraisal of the situation after examining the showing. It was a poverty stricken patch of mineralization—a fraud—but as far as the promoters were concerned it was a cash cow that yielded personal wealth at the expense of credulous investors.

The third frame we entered was the cook tent. It still had its canvas roof. Inside we found a collection of empty coffee cans, jam and pickle jars, and baking powder canisters. In the corner, I found a teaspoon and dropped it into my pocket. This would go to our cook. We also found a table that had only two front legs, the rear being nailed into the tent frame. A few blows of the geology hammer loosened it, and it was deposited outside so it could be carried to the chopper as soon as scrounging was completed.

We had visions of the chopper mechanic adding two rear legs during the day, while we were traversing, and making a stand in front of our tent with a bar of soap and a wash basin on it to ease the shock of awakening at 6:30 in the morning. Across the end of the tent, I found a 6 foot length of two-by-two lumber that had served as a pot and pan rack. I extracted most of the nails on which they had hung and laid it beside the two-legged table. It would be used to make the missing legs.

The next frame we entered was probably the bunkhouse of the prospectors and their field assistants. We found nothing of interest. The drill crew used the next tent. It contained an assortment of broken gears that were still greasy and a litter of chain links, short lengths of frayed cable, a couple of empty grease pails, and a pile of black rags. At the back of the building was a pile of empty oil drums. I collected three bungs that fell into my pocket. We were always losing them.

We also noted an eight foot, two-by-six plank in excellent condition. A plank of this size and strength is necessary for rolling gas drums out of float planes, so that loaded drums did not drop straight down and crunch a float. This is the closest thing to disaster that could befall a camp that was wholly supplied by air. We noted it, but did not move it. It was too heavy to be carried by the chopper, and there were more important items to acquire.

The last tent frame was the warehouse. It was beside the lake at the only place for a considerable distance where deep water extended inshore. From its door a sturdy dock extended into the lake. The warehouse's fiberboard sides had been bashed down to facilitate emptying its contents when camp had been abandoned. Strewn outside, where they had been flung, were half a dozen bundles of unused core boxes still strapped together in packets of six. We thought of taking a couple because they made excellent shelves when turned upside down, but we left them. There were also several bundles of four foot wooden pickets that were used to lay out ground grids to facilitate detailed mapping of mineralized zones. We ignored them. Behind the warehouse were twenty-five partially stacked, empty gas drums waiting to be flown out. This would never happen.

We collected our stools, tables, and lumber, and stacked them by the chopper and paused to study how they could be lashed aboard so they would present minimum wind resistance. Most of the items were light but bulky. As we surveyed the situation, our eyes caught sight of two aerial masts on the ground. They were of superb construction, made of number one, straight grained lumber. The masts were A-frames with poles bolted into their apexes. They were about five meters tall and between them was thirty meters of tangled copper aerial wire. This was a real discovery. Base camp's radio would be without an aerial at the treeless Tree River camp, and our crew would be without the music of CBC, BBC, Voice of America, and Radio Moscow.

The more we looked at them, the more determined we were to have them. We twisted off the rusty haywire that anchored them to heavy boulders and dragged them to the chopper. I wound the aerial wire into a coil and went looking for the lead wire. While I was looking, Fred got a wrench and pliers from the chopper's toolbox and dismantled the A-frames. He laid the lumber of the A-frame on the carrying rack on his side to see if they would fit without blocking the door so he could get out in an emergency. They fit. I found twenty meters of twisted lead wire behind the nearest tent frame, which I wound into another coil and laid under the stools, tables, and the two 10 gallon gas drums tied onto the rack on my side of the chopper. Every time we made a long traverse, we took twenty gallons of gas in ten gallon drums and refueled as soon as the chopper's tanks would hold them. We always brought the empties back.

When Fred had lashed the dismantled A-frame masts to his side and I had tied the other items to my side, he inspected the load and decided it was not quite balanced, so he added a couple of pieces of lumber to my side. After the load was secured, we made one more tour of the tent frames to see if we could find something else of small size that might add to our comfort. We found nothing. When we got back to the chopper, Fred started the engine and, while it was warming up, I dug the steel ball from my pocket, went into a pitcher's windup and threw it at the fiberboard walls of the nearest tent frame, fifteen meters away. It punched a neat round hole, and we headed for the Tree River camp.

—

CHAPTER 6

Dam Country

We arrived at the Tree River camp at seven o'clock in the evening.
It was an hour later than planned, but the delay was not completely
due to our pause to scrounge. In the middle of the Tree River watershed, while flying over some of the most spectacular scenery in
North America, we found a gossan and, because there was a good
landing site nearby, we examined it instead of noting it on the map
and visiting it at a later date. Besides, we had been flying for an hour
and a half, and it was time to take a break from our cramped seats
in the bubble.

The gossan was adjacent to a thick diabase dike and I applied
the field test to see if it contained nickel minerals. The test was positive. The showing was small and the nickel mineralization was
associated with the diabase dike. These occurrences are of little economic interest, but this one indicated that nickel mineralization
might be present in related rocks, if any related rocks were present.
I spent about twenty minutes examining the gossan and locating it
very accurately on the map, and another ten minutes flying around
the immediate area. I found several additional rust patches, but
because our labor was needed at base camp, we left the rest of the
region to be prospected another day. To reach camp, we had to fly
the length of the middle portion of the Tree River and this is the
region I will describe.

Base camp was located five kilometers upriver (south) from the
mouth of the Tree River (where it empties into Port Epworth). Port
Epworth is 110 kilometers north of the Arctic Circle, and at this latitude the continental glacier moved northward. Port Epworth is a
small, bottle-shaped inlet that protects the Tree River estuary from
swells generated on Coronation Gulf by northern gales. The Tree
River is about 130 kilometers long, but there are no trees in its entire
watershed. In fact, there are no trees within seventy-five kilometers
of its headwaters, although, some thickets of arctic willow are found
in more sheltered parts of its valley. Arctic willows, however, are
not trees.

For fifteen kilometers above tidewater, the Tree River flows in

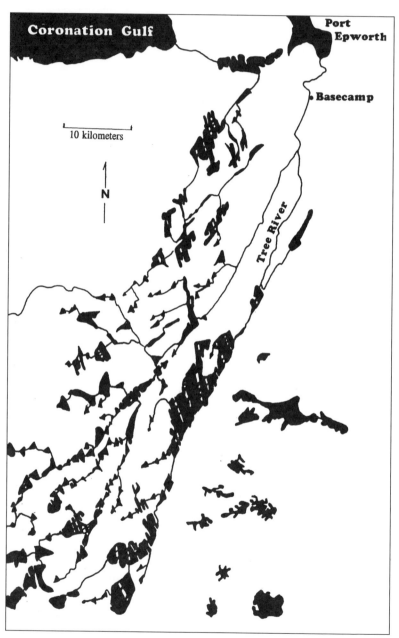

Coronation Gulf

Port Epworth

Basecamp

10 kilometers

N

Tree River

Dam Country on the Tree River

a flat, wide valley. The river itself is narrow, slow-moving, relatively deep, and has a sandy bottom. It is also relatively straight. Base camp was located just upstream from the first major bend, where the water was fresh and where there was protection from waves generated by northern winds. Fresh water is essential for drinking and cooking. It also protects the plane, because salt water spray during takeoffs and landings is highly corrosive of the plane's aluminum body and engine.

Our campsite was at the edge of a sand dune that had its base anchored by a carpet of sedges and mosses, but its crest was loose and moved during high winds. Fortunately, the sand was coarse and it did not infiltrate our tents while we were there. As a precautionary measure, however, the chopper and plane mechanics always kept polyethylene covers over the engines of their craft when they were on the ground.

The sand dunes were elongated in a north-south direction. The sand had originally been laid down in the shallow waters of Port Epworth during the terminal melting stage of the continental glacier, when the Tree River channel was a major drainway. When the glacier was at its maximum extent, the enormous weight of the ice (an estimated 2,500 meters near the mouth of the Tree River) had depressed the earth's crust by 200 meters, in the same way a person depresses the cushion of an automobile seat when he sits on it. After the ice melted, the land rebounded. The river's delta emerged as a sand plain, and strong winds have removed the fine sand and heaped the coarse sand into dunes.

The middle portion of the Tree River, which begins about twenty kilometers south of base camp, has some of the most spectacular scenery in North America. The scenic area is twelve kilometers wide and approximately ninety kilometers long, and is immediately west of the river. I had heard it described several times before I made the traverse in June to find open water. On that traverse, I had seen its outlines but there was still too much snow to see its grandeur.

The eastern side of the Tree River valley is a steep, continuous escarpment, composed of granite, but the western side is nearly flat for a distance of fifteen kilometers until it merges with the interior plateau. The gradient of the Tree River, from the inland plateau to the raised delta where base camp was located, is fairly steep. Rocks exposed in the valley are essentially two types: a thick sequence of folded, multicolored limestones, and a swarm of diabase dikes and sills. For eighty kilometers, the dikes and limestones are 100 percent exposed in the Tree River valley, because they were scoured clean by an avalanche of meltwater that poured down the Tree River drainway during the final stages of melting of the last continental glacier. For two or three months during the few years of greatest runoff, the Tree River drainway was up to twelve kilometers wide and was a roiling flood of cascading water up to thirty meters deep.

Diabase dike swarms form during periods of continental collision and continental rifting. A collision of continental plates is the probable cause of the dike swarm exposed in the Tree River drainway. Although, there is no good evidence of plate collision at the time the dikes were emplaced (1.27 billion years ago). During continental collision, enormous compression produces mountain ranges, like the Swiss Alps, where the Italian mini-plate collided with southern Europe. When the compression relaxes, fractures open in the rigid continental crust and the underlying lithosphere. The semi-rigid lithosphere is 140 kilometers thick and, with the continental crust, forms the outer shell of the earth.

Below the lithosphere is the asthenosphere (outer mantle). Seismic data indicates it is about 700 kilometers thick. Its uppermost layer is composed of partially melted basalt. Whenever the outer mantle is tapped by fractures that extend through the lithosphere, the reduction in pressure causes the partially melted basalt to liquefy into magma. Magma has a lower density than the cool rocks of the lithosphere and it rises to the surface where it flows out as lava. Lava flowing from these fractures is called flood basalt. Flood basalts are very common. In Ontario, they are found in small patches along the eastern shore of Lake Superior. In the eastern United States, they are present in large quantities in northern Michigan, northern Minnesota, and on Isle Royal. In the western Arctic, they are found near the big bend of the Coppermine River, at Bathurst Inlet, and on Victoria Island.

Flood basalt magmas are quiet effusions. They flood into valleys and fill them, then overtop the hills. Hills over 1,000 meters high can be buried beneath a pile of horizontal lava flows until the pre-flood basalt topography is hidden under a high plateau. The dikes exposed in the Tree River valley are the roots of a plateau of flood basalt that was once continuous, with the basalt lava flows preserved in the Coppermine plateau 130 kilometers to the west, and at Bathurst Inlet 150 kilometers to the east. Erosion has removed the lava flows at Tree River, and exposed the fractures through which the magma reached the surface. These fractures are filled with basalt that solidified before reaching the surface. The basalt that slowly solidified in fractures has a distinctive texture composed of a felt of interlocking needle shaped crystals. It is called diabase. The texture of diabase dikes makes them very resistant to glacial abrasion so that, when they intrude softer sediments, a moving glacier sculpts them into linear ridges, if they are parallel to the moving ice.

On its way to the surface, the magma passed through a thick pile of pastel colored limestones. These rocks are soft shades of red, green, gray, buff, cream, and brown; with a few thin beds of bright red, white, and black; but shades of buff and cream make up most of the exposed rock. Weathering had muted the strong colors that are found on freshly broken surfaces. For example, a rock that has a strong bluish tinge on a freshly broken surface will weather to buff or cream, and one that is crimson on a fresh surface will soften to pink on its weathered outcrop. From the air, the outcrop pattern of these limestones is similar to a finger painting of regular and irregular swirls of different colored paints. Imposed on the folded limestone beds is a sharply angular pattern of straight black lines made by diabase dikes and sills. This, in turn, is overlain by a large number of intensely blue triangular shaped lakes.

Near the river the folds in the limestones are open and have the color definition of a cameo. To the west, however, the colored limestones form jagged patterns, like glass fragments in a kaleidoscope, because as one goes west from the river the folds become more compressed. Beds that dip twenty-five degrees near the river, dip seventy degrees or are vertical twelve kilometers to the west. The more deformed rocks to the west are also broken by a complex system of faults. The whole western fringe of the valley is a jumble of sawtooth ridges and valleys that are only two or three meters deep, until the ground moraines of the interior uplands bury them.

Cutting across the colored limestones is a swarm of diabase dikes. They intersect the course of the Tree River at a long angle. A good estimate of the frequency of the dikes can be measured by the number that are more than 100 meters thick. Dikes this size occur about every kilometer, and between them there are as many as fifty dikes ranging in thickness from one to fifty meters. The dikes outcrop as straight black lines and everywhere they form elongate ridges with steep sides. They cut across all colors and create an extremely distinctive drainage pattern. In the larger lakes in the Tree River valley, the thinner dikes form lineal chains of black islands or pencil shaped peninsulas. Even when they are underwater, thin black dikes can be traced as dashed shadows through the centers of lakes.

At this place in the Arctic, the continental glacier was moving northwards and the dike swarm was thirty degrees across its path. As the ice moved, it tended to slide along the length of each dike and shape them into linear ridges. At the same time, the glacier gouged out the softer limestones on the upstream side of each dike, which created deep basins behind the widest dikes. The wider dikes were sculpted into natural dams. The basins behind these dams are now lakes. The shores of these lakes converge in an upstream direction to form triangular shaped lakes. The converging shores are composed of multi-colored limestones, but the downstream shores are ruler straight and black.

The natural dams on the Tree River are a textbook example of how differential erosion can produce a startling surface feature. The geometric shape of the lakes and the continuously elevated black lines of the natural dams focus attention on a world class example of rock sculpturing by glaciers. The river drops over the dams as waterfalls or precipitous rapids. Some waterfalls are over twelve meters high, and others drop only one meter. These natural dams have impounded the Tree River into a series of lakes that descend to ocean level like a staircase. Someday this geologically spectacular region will be accessible to tourists and, hopefully, it will be a national park.

CHAPTER 7

Home on the Range

We spent our first day at Tree River putting the camp in order and resting. However, the two aircraft continued to fly. The plane made two trips to Shoestring Lake to bring the remaining supplies and Keith used the chopper for an exploration traverse. In the morning, he revisited the gossan I had found the previous day and thoroughly examined the dike swarm looking for similar occurrences. He found several small gossans in its general vicinity but nothing exciting. He spent the rest of the day examining the limestones in the Tree River valley and the interior plateau. Keith found several clusters of gossans in schist beds among the limestone beds. Where he was able to sample them, they contained a few specks of copper sulfides. It was a mildly encouraging discovery, indicating that the Tree River limestones might host a sedimentary copper deposit.

The gossans were spread over several square kilometers and seemed to be confined to a single bed, but the complexity of folding made it difficult to determine if this was actually the case. The persistence of mineralization, however, indicated the possibility of a sedimentary bed with large amounts of biogenic sulfides that could contain a concentration of copper minerals. In small marine basins in areas of volcanic activity, copper sulfides are often precipitated from seawater by bacteria. The most common biogenic sulfide mineral is pyrite, but in several areas in the world copper sulfides predominate. Biogenic copper deposits in sedimentary rocks are among the largest and richest in the world.

All of the lakes in the Tree River country were ice free, apparently having been blown open by the same gale that cleared Shoestring Lake. The day after camp was established, the junior geologist and his assistant were flown out to a fly-camp in the area of the nickel gossan, and my assistant and I were placed in the area of sedimentary gossans. Shaver dropped off the junior geologist and his assistant, and went on to Yellowknife for a load of aviation gas. A couple of hours later, I was taken to my area on the return leg of a charter flight that brought gas from Yellowknife. We took the canoe because several large lakes in the prospecting area could be accessed by short portages.

When we were ready to leave, the plane was turned around so the rear of the floats were grounded on sand and the nose pointed toward the middle of the river, but the engine coughed and spit for nearly five minutes before starting. The pilot had told us that this particular engine was newly rebuilt, and new or rebuilt engines are always the ones that give trouble. We felt an unexpressed concern. To our perked ears, the engine performed perfectly. However, the pilot noticed trouble on the gauges, which he kept to himself until we were on the water again. Before setting down on the lake where our camp would be located, we circled the area to get a mind's eye picture of the terrain, particularly the location of the gossan zones nearest lake shores. Collecting assay samples by canoe traverses would allow us to gather the maximum amount of information in the shortest time.

We landed on a large centrally located lake, shaped like a long arrowhead. It was at the western edge of the dam country. Our campsite, like all of the other nearby lakes, had no sand beach. This complicated unloading. I chose a campsite at the north end of the lake where there was some protection from high waves generated by north winds. We unloaded the plane on a low rock peninsula where there was deep water inshore. The low shoreline allowed the plane's wings to swing over the peninsula while it was being unloaded without hitting rocks.

While we were unloading, the pilot searched for the cause of the trouble he had noticed on the gauges. The trouble was gas consumption. The trip should have consumed six gallons of gas, but it used eleven. When he unscrewed the side cowling, he found that the fuel line was loose due to careless tightening. The new fuel line had been stiff when it was installed and, on the last quarter turn, it had failed to lock into place. One-third of the gas coming through the pump had spilled near the hot engine and had been blown out by the cooling air coming through the cowling.

We had been lucky to avoid a fire. He fixed the gas line and, as soon as we were unloaded, he made ready to leave. We turned the plane around and shoved it toward the center of the lake, but we did not give it enough momentum for the pilot to steer clear of a rock shoal. He had to paddle the plane into a safer position in the middle of the lake before starting the engine.

But, the engine would not start. Before he was out of sight or in danger of drifting onto the far shore, he got out on a float and gestured and yelled across the lake. We stopped setting up camp and got into the canoe to see what was the matter. He described his troubles. Every two minutes he had tried to start the engine by turning on the booster pump, using the primer, and activating the spark booster coil, but only one or two cylinders would fire. Round and round the prop had gone, occasionally sputtering but not catching. He had done this for fifteen minutes before getting out of the cockpit and calling for help.

We guided the plane to a safe place along the rocky shore, but, in order to keep it safe, my assistant held the tail to prevent it from drifting, and at the same time he pushed the plane away from contact with the shore. I helped the pilot work on the engine. He could have both hands free by entwining one foot around a rung of the ladder to the cockpit while draping himself forward over the engine. Again he unscrewed the cowling and looked. While he had been drifting, he had been mentally calculating what could be wrong. His main suspicions fell on the primer so, while he searched, I stroked it a couple of times. He found the trouble.

The function of the primer is to push an extra charge of gas into each cylinder to aid ignition. The priming tube has nine holes in it with a small tube leading to each cylinder. There are always more holes than cylinders but the extra ones are always plugged with screws that are locked in place so they cannot work loose. All the extra holes had been plugged during overhaul but one had not been safetyed and this one had worked loose. Although this hole was only eight millimeters in diameter, it allowed enough extra air to be sucked into each cylinder to dilute the gas mixture below the point of ignition. The pilot began searching for a substitute plug. I told him I had no gum, but I volunteered a pencil top eraser. This was not very satisfactory so, on the off-chance that the plug might be caught in the cowling sump, he opened it and searched. By a merciful act of God it was there, and fifteen minutes later it was in place and safetyed-in. The engine started with hardly a prime and he took off.

Although the relief of our immediate area was no more than eight meters, the ground was extremely broken. Most of the sediments were resistant limestones and diabase dikes that formed ridges, but there were several beds of soft schists that the glacier had excavated into narrow trenches. These were difficult to cross.

Faulting complicated the pattern by making the ridges and valley discontinuous, so that we could not get on a ridge and follow it for more than half a kilometer before having to drop into a blind valley that often contained a small lake. The gossans we were looking for were in soft schists, but we were not sure whether there was one bed or several.

We had two objectives on our first day's traverse. The first was to locate and examine the nearest gossans to see what kind of mineralization it contained. We hoped it would be associated with a distinctively colored limestone bed that we could use as a marker. If we found that a gossan was always above or below a distinctively colored limestone bed, we could predict that mineralization was present, even though it was not exposed. Secondly, we wanted to get some idea of the geologic structures we would encounter in the short time we would be in the area. Long traverses would be difficult. Most traverses would have to be made by taking the canoe to places along the shore and walking inland from there.

In the morning, we took the canoe down the lake and walked a kilometer inland to examine a small group of gossans. We found only a few specks of copper minerals in the predominant pyrite mineralization, but we did find a distinctively colored limestone bed above the gossan that would serve as a marker to indicate the presence of the pyrite bed at other localities. In the afternoon, we prospected the lakeshore. The shoreline was too long to be prospected in one afternoon, but during the first day we gained a good idea of the structure and mineralization we could expect.

While paddling home that afternoon, I made the decision that if we had good weather in the morning we would make a long traverse to the west to look at the most distant cluster of gossans that Keith had located from the air. If, afterward, the weather turned bad, as was likely, and remained that way for an extended period, we could do shore prospecting and examine the nearest gossans during short periods of good weather. It was on this long traverse that we encountered a herd of muskoxen. We knew some were in the area because we had seen a herd when the plane set us in.

For a long time muskoxen were not authoritatively classified in any mammalian family. It had originally been classified as a transition group between the ox and the sheep, and was named Ovibus moschatus, which it still retains, but a definitive anatomical study showed that muskoxen belong to a group in transition between the ox and the goat. Its only living relative is found in the cold rain forests and rhododendron thickets of the middle elevations of the Himalayan Mountains in northern India, southern China, and Tibet. Its range does not extend into the higher, harsher climates of the Tibetan plateau where the yak lives.

The greatly reduced herds of caribou and muskox are the last survivors of millions of large mammals that once grazed the tundra from northern Europe to eastern North America during the four interglacial eras of the past 500,000 years. During the four eras of continental glaciation, the Bering Sea was a dry plain that afforded easy east-west migration of animals from Asia to North America and North America to Asia. The Bering Sea was emergent because the continental glaciers of Europe, Siberia, and North America removed enough water from the oceans to lower ocean levels by about 120 meters. During the great melting of the last continental glaciers, that began about 25,000 to 30,000 years ago, humans entered North America while hunting large mammals grazing on the Bering plain.

When humans entered the interior of North America about 10,000 years ago, a close relative of the muskox was living as far south as Arkansas, and competed for existence with the mammoths and the American mastodon, great bison, camels, and horses. All of these species are now extinct, most of them so recently that arrowheads and charcoal from cooking fires have been found associated with their bones. The endangered muskox is now reduced to about 3,500 animals living in three isolated areas in the barren lands: Nunivak Island (off the coast of Alaska), Ellesmere Island (the northernmost of the arctic islands), and in a broad belt on the mainland from Tree River east to Hudson Bay. Perhaps 20 per cent of the surviving animals live between the Tree River and Bathurst Inlet. In these remote regions, there are few human predators.

The muskox is not a large animal when compared to the bison or horse but, when its environment is taken into consideration, it is a huge animal. It is the largest herbivore inhabiting the Arctic, and is exceeded in size only by the polar bear among arctic land animals. Mature animals are about one and a half meters high at the shoulders and weigh between 180 and 350 kilos. It lives on the short grasses, sedges, mosses, and lichens that grow in the rockiest parts of the barrens. In winter, it finds food on the hilltops where the wind

Muskox

blows away most of the snow and where it can scrape off the thin cover that remains. Vegetation growing on this seemingly sterile ground is very rich in nutrients, but even so muskox lose most of their fat during the winter.

Muskox have two very distinctive characteristics: their wool and horns. The guard hairs of its fur are more than thirty centimeters long and form a shaggy skirt that hangs almost to the ground. Its outer guard hairs are dark brown to black when prime. Under the guard hair is a thick, dense layer of fine wool that is one of the lightest and best insulating fibers known. During the summer it sheds some of its light colored undercoat that twists into streaks of pale wool that hang from its sides and back, giving the animal a mangy and mottled appearance.

Both sexes have horns. The horns form a massive plate across the skull and are sharply hooked and pointed. The head is carried close to the ground while the shoulders form a high, bison-like hump. A quick toss of the head would catch the belly of a wolf trying to attack it from the front, and a twist of its neck would disembowel it and flip it off to one side. Its hooves are another defensive weapon. They splay outward and give muskoxen a large surface for walking on snow, but in the summer such wide hooves on stubby legs give it a grotesque appearance.

Muskox are always in a herd except for the occasional old bull, but he is usually not far from a herd. Although herds are never large, perhaps eight to twenty, they form an effective defensive unit. When a predator approaches, the animals move very close together, flank to flank, and present the enemy with a wall of low, hooked, horns. The calves hide close behind. If the herd is attacked from more than one direction, they will form a defensive circle. On one of the few occasions I saw muskox on the ground, I witnessed the defensive strength of the herd against a wolf pack. I had heard how the wolf is the deadly enemy of the caribou and when a wolf wants a specific caribou in a herd, it usually gets it by the teamwork of the pack. It was my good fortune to see a confrontation between muskox and wolves.

The most distant gossan cluster we had to sample was along the north shore of the next large lake to the west of our fly-camp. It was about nine kilometers away. While circling in the plane prior to setting up camp, I had noted that there was a low moraine hill between Arrowhead Lake (the one we were camped on), and the lake where we were going. I wanted to cross the moraine at its widest point because its relatively smooth surface made walking much easier than the broken terrain of the ridges and valleys. While we were in the maze of ridges and valleys, we prospected, but because the day's traverse was long we needed some easy walking in order to have time to examine the gossans when we reached them. By eleven o'clock in the morning, we had covered six kilometers of broken country and were just coming to the moraine hill. I was debating whether we should have lunch on the rock outcrops where we could find a dry seat out of the wind, or walk across the moraine and have lunch on the other side. We did not have to make this decision.

As we came to the top of the last rock outcrop before the easy walking on the moraine, we saw a small herd of muskoxen about 500 meters away. It is easy to see this distance in the barren lands when you are on a hill. We were downwind and had not been detected. We wanted to get closer and we saw how we could do it. Immediately behind the rock ridge we had just climbed, we had crossed a trench valley that angled in the direction of the muskox herd. It would afford us complete concealment for a much closer observation. We backed off the ridge and walked to the end of the valley and crept up and over the rock ridge. The herd was on the crest of a low hill about 150 meters away. There were five adults in a semi-circle facing south, with two calves in the rear.

Our curiosity was unbounded when we saw that they did not move and nothing around them moved, and they were not feeding. We waited twenty minutes in silence but the only movement was an occasional uncoordinated shuffle from side to side. We began talking in a low voice, speculating about the cause of their immobility, then we began eating lunch. It had no effect on them. When we had finished eating we decided to make a cautious approach and see what we could see. We thought their posture was probably due to wolves that we did not want to encounter because we had only short knives and our geology hammers for weapons. But, we were even more curious to know why they were immobile.

We were about to leave our lunch table outcrop when the herd made our approach unnecessary. After an ungainly, uncoordinated shuffle, with lots of pushing and shoving, two animals of the herd completed the defensive circle. Almost immediately we saw two gray-white wolves in the rocks between the muskoxen and us. They had been trying to get around behind the semi-circle and attack the calves but had failed and were now further downhill from the herd and at a tactical disadvantage. Since they were discovered, they moved around in the open, running here and there, back and forth, but not venturing too close. The other three adult members of the herd faced north so we presumed there were other wolves concealed among the rocks, although we did not see them.

The muskoxen remained in place with an occasional side to side shift of their heads. If a circling wolf approached too closely, a hoof was stamped. Their eyes were always on the wolves. Occasionally, a wolf would come close to the nose of an ox but no further. It was not prepared to spring forward but to jump back out of the way of a hoof or horn.

We watched them for half-an-hour and the only time they changed positions was when they swung around to form a full circle. If muskox survival depends on watchful waiting, they could outwait the wolves or, if the wolves attacked, they would probably have been the ones to suffer a loss. We had no desire to disrupt natural selection by trying to scare the wolves away. If we had tried, our greatest danger would have been the hooves and horns of the muskoxen, if they charged and we lost our footing. They would have stomped us flat. We were content to be momentary observers while doing our job.

CHAPTER 8

Chocolate Fog

High summer arrived on the arctic coast shortly after our encounter with the muskoxen. We were bathed in warm, southerly winds from a continental high that was blowing up the length of North America. We had no fog, nor overcast skies and, on July 20th, the air temperatures reached the low thirties. We stripped to our t-shirts for field work, which we could do because the wind kept the flies away. It blew continuously with gusts up to thirty kilometers and pushed widely spaced altocumulus clouds northwards in continuous columns. The sun was rarely obscured and, by two in the afternoon, the darker rocks were too hot to sit on.

The traverses we ran during these days confirmed the stratigraphic location of the pyrite (iron sulfide) bed and led to the discovery of additional outcrops, most of which were small, and none of which contained any quantity of copper minerals. Only assays would tell if the pyrite contained gold or silver values. The pyrite was in soft schist that was a low grade iron formation containing about 15 percent iron. It did not usually outcrop but when it did it was highly visible because the pyrite oxidized to ocherous red hematite.

The sixty meters of limestone immediately above the pyrite iron formation contained 3 to 5 percent rhodochrosite (manganese carbonate), alternating with thin beds of siderite (iron carbonate). Rhodochrosite is a pink mineral, but when it oxidizes it produces a mineral that is coal black and, when siderite oxidizes, it produces red or brown oxide minerals. Both the pyrite iron formation and the carbonate beds were weathered to a depth of one centimeter, and some of the black stain from oxidized rhodochrosite had migrated down the sides of the ridge and mixed with the red stain to produce a layer of brown dust. The black stained limestone made a highly visible marker bed that told me the pyrite iron formation was next to it.

Surprisingly, carbonate rocks weather rather rapidly in the Arctic because the generally cool weather slows evaporation of rain so that limestone has more time to go into solution. Solution weathering converts the iron and manganese carbonate minerals into oxides. Oxides stay in place, as residual minerals, because they are nearly insoluble in water. Oxide minerals form a crust that is held together by water during the summer and cemented by ice during the winter, but when it completely dries, which is not often, it becomes a light powder. In some places, a person walking on the oxide crust leaves a footprint.

On July 22nd, the weather changed. We got up anticipating another day of good weather, and were in the field by 9:30 a.m. We had planned to take the canoe to the south end of Arrowhead Lake and portage fifty meters to the next large lake, go to its southern end, then walk five kilometers to the east, and examine a gossan cluster that Keith had found. We had already mapped the shoreline of the southern lake and had found an extensive, but hidden, outcrop of the pyrite iron formation. Keith had missed the outcrop in his aerial survey because he passed on the wrong side of the peninsula where it was located, or because the sun was in his eyes and the glare obscured its color. By the time we grounded the canoe at the southern end of the lake, the wind had died and wispy streaks of low overcast began coming from the north.

The brilliant glare of the sun became diffused and muted, and the air grew cooler. At this point, I decided to have a leisurely lunch and see what happened with the weather. Just after lunch the wind began blowing strongly from the north, and the overcast dropped lower and became thicker. We thought that fog or misty drizzle was in the offing, so we ended the traverse and began the return trip to camp. If the low overcast changed to fog, we could follow the shoreline to camp. Fog would not worry us. Before we got started the fog arrived but, instead of lazily enveloping us, it was propelled by a ninety kilometer gale. It was cool and dripping, and we were soon chilled as we huddled behind a rock waiting for the wind to abate. We were giving ourselves to self-pity, trying to justify our misery, when a quirk of nature made us temporarily forget our discomfort.

The fog seemed opaque. We commented to each other that it must be getting thicker, but then we noted that some gusts, as they eddied around us, seemed to be slightly darker. To check this, I took off my glasses because they had become covered with small beads of moisture. There was a brownish coat on the lenses, like a fine paint spray, and my assistant noticed that my eye sockets were whiter than the rest of my face. He wiped his face and it streaked. I did the same and mine streaked. We scrambled to the top of the outcrop and looked into the fog. As far as we could tell, the fog was brown. We thought we saw streaks in it, some darker than others, we thought, but we could not be sure. The fog got darker, and we could no longer doubt that it was brown. As we sat trying to imagine an explanation, the wind began to drop and our minds turned to thoughts of dry clothing and a warm tent.

Even though the wind was still blowing thirty kilometers in our faces, and the waves were nearly half a meter high, we launched the canoe and started for home. We calculated that, if we could turn the gossan point, we could avoid the worst of the waves by staying close inshore to take advantage of the lea it afforded. By the time we turned the point, we had both put forward the obvious explanation.

The high winds of the advancing front had picked up iron and manganese oxide dust that formed a crust on the carbonate rocks and gossans. The red, brown, and black dusts were mixed into a chocolate powder that was blown into the air and held in suspension by the fog. The fog coated everything it landed on with a metallic sheen, especially our skins. When we got back to the tent, we found that our faces glistened like new cars. We had been exposed to the chocolate fog for about ten minutes before water droplets wetted the gossans surface and stopped dust from being blown into the air and mixing with the fog. We had experienced a very momentary event by being in the right place at the right time.

After the front passed, the temperature dropped to nine or ten degrees, and the wind diminished to a steady twenty kilometers. For the next two days, the winds continued at that velocity, and fog was alternately present as a thick mantle or thin veil. But, the winds never dropped low enough, nor did the fog lift long enough, for us to do any more prospecting. Calm weather returned early in the morning of the fourth day and the plane came and got us.

CHAPTER 9

The Muskox Intrusion

The plane did not return directly to Tree River camp. Keith was aboard, and we flew west to the Muskox Intrusion where INCO had a diamond drill in operation. We had to log the core and bag it for assay. This was a three man job if it was to be done in a day. Keith also needed our canoe. The drill was eighteen kilometers north of the Desolation Lake camp, on the shore of a shallow, rock-filled lake on which the plane could not land. The plane landed three kilometers away and, by making three short portages, we could paddle the canoe on a series of shallow lakes to within 100 meters of the drill.

The Muskox igneous event occurred 1.27 billion years ago. Its age has been dated by radiometric techniques that measure the rates of decay of several radioactive isotopes. When igneous rocks crystallize, many minerals incorporate very small amounts of radioactive elements into their crystal lattices. The radioactive elements have the same isotopic ratios found in the earth's outer mantle (asthenosphere), where the magma originated. After magma crystallizes, the radioactive isotopes of these elements decay at a constant rate (a predictable half-life). Measuring the ratios of these isotopes in minerals where they concentrate, dates the age when the magma was molten and emplaced. The two most common radiometric techniques for measuring the ages of igneous rocks are: potassium/argon and uranium/lead.

Potassium is a component of feldspar, a common rock forming mineral. One of its isotopes, potassium-40, decays to argon, which is an inert gas that does not combine with any other element. Argon remains trapped within the lattices of minerals containing potassium. The only source of argon is the decay of potassium-40, and it takes 1.3 billion years for half of potassium-40 to decay to argon. It is possible to calculate the age of igneous rocks by separating feldspar and other minerals containing potassium from the other minerals in a rock and measuring the ratio between potassium-40 and argon.

The second way to date igneous rocks is by the uranium-lead

method. Very small amounts of uranium are present in the lattices of many minerals that are present in igneous rocks. Zircon is the usual mineral chosen for analysis, because it is a common accessory mineral and is easy to separate from other rock forming minerals. Most uranium is isotope 238, but a second isotope, in much smaller quantities, is uranium-235. It has a half-life of 713 million years, and its decay product is lead-207. All of lead-207 in igneous rocks is derived from the decay of uranium-235. By calculating the ratio between uranium-235 and lead-207, it is possible to measure when an igneous rock was molten and emplaced in the earth's crust. Since both potassium-40 and uranium-235 have slow rates of decay (long half-lives), they are well suited for measuring the ages of ancient rocks.

The lithosphere is the outermost concentric shell of the earth. On average, it is 140 kilometers thick beneath continents, where the rigid rocks of continents form the uppermost forty kilometers. Under oceans, however, the rigid crust averages only seven kilometers in thickness.

The asthenosphere is the concentric shell directly beneath the lithosphere. Seismic data indicates that it is 700 kilometers thick. The upper 300 kilometers is in a state of partial melting, at a temperature of 1,200 to 1,400 degrees. Seismic waves, generated by deep earthquakes, indicate that the rock in the zone of partial melting has the composition of basalt. The magma that was the raw material for the Muskox Intrusion originated in the zone of partial melting. Basalt is a rock that concentrates the metallic elements of iron, magnesium, chromium, and titanium, and, compared to continental rocks, has a high concentration of nickel, copper, and platinum.

As long as the layer of partial melting in the asthenosphere is at equilibrium (under gravity pressure), the metallic atoms of nickel, copper, chromium, iron, and platinum are not tightly bound to other atoms. If fractures pierce the lithosphere, basalt in the zone of partial melting liquefies, moves toward the fracture, and ascends to the surface of the earth as magma. As the magma ascends, both temperature and pressure decrease, and nickel, copper, and iron atoms seek to chemically combine into minerals. If the ascending magma contains large amounts of sulfur, or acquires large amounts from the rocks it passes through on its way to the surface, these elements have a high affinity for sulfur and combine to form sulfide minerals.

Sometimes the basalt magma does not reach the surface as lava flows. It is emplaced in reservoirs within the crust. This is the origin of the Muskox Intrusion. It was emplaced in a reservoir that was seven to nine kilometers below the surface. After emplacement, the Muskox Intrusion slowly cooled. In large igneous intrusions that slowly cool, like the Muskox Intrusion, gravity differentiates minerals into layers with the heaviest minerals at the base and the lighter minerals at the top. Sulfide minerals begin the crystallization process. They settle downward because they are much heavier than the remaining magma. They form a basal layer that concentrates the nickel and copper in the magma. This is the type of mineral deposit that INCO was exploring for in the Arctic.

Four other metallic elements in the magma have strong affinities for oxygen and crystallize as oxides. They are iron, chrome, titanium, and vanadium. In large igneous intrusions, they often form gravity differentiated layers. Only iron forms sulfide, oxide, and silicate minerals. In magmas that never reach the surface and slowly cool in large reservoirs, oxide minerals crystallize later than sulfide minerals, and form layers higher in the magma reservoir than sulfide minerals. Oxide minerals must have the right temperature, pressure, and availability of oxygen to precipitate from the melt and this does not usually occur until a high percentage of silicate minerals with high iron and magnesium contents have crystallized.

Some oxide minerals act as sponges for other metals: magnetite (iron) is usually enriched in titanium; ilmenite (titanium mineral) is always enriched in vanadium; and chromite (chromium mineral) is enriched in iron and aluminum, and often precipitates with platinum and the last remaining sulfide minerals in the magma. When these minerals concentrate in layers within a slowly crystallizing magma, they can become ores of platinum, chromium, titanium, or vanadium.

In 1956, just after the discovery of the Thompson orebody in Manitoba, INCO's senior exploration geologist discovered the gossans that cap the nickel-copper sulfide mineralization in the basal zone of the Muskox Intrusion. INCO's focus of exploration in 1957 and 1958 was the basal zone of nickel-copper mineralization because of its potential for making a mine. By 1959, enough drilling had been done so that it was clear that the basal zone of nickel-copper sulfides was not large enough, nor of sufficient grade, to make a mine. Company geologists had also recognized that concentrations of platinum minerals might be present, if there was a layer of chromite higher in the intrusion.

The potential for a platinum deposit, however, was untested. In 1959, a prospecting team found thin chromite layers at two locations, but they were low grade and poorly exposed. There remained the possibility that ground moraines might conceal a thicker and richer layer of chromite that contained disseminated platinum-nickel-copper minerals, similar to what was being mined in South Africa. In mid-June 1960, when snow still covered 80 percent of the hills, a crew was flown north to drill two holes at widely separated localities to intersect the zone where the thin chromite layers had been found. Keith had located the drill sites, and we were returning to log the core of the second hole and bag it for assay.

In the spring, there were modest hopes of finding a platinum layer thick enough to mine, but assays from the first hole indicated that there were only two thin chromite layers and they contained little platinum. The highest value in the thickest chromite layer was one and a half grams platinum per ton over a thickness of ten centimeters. This is very low grade. At this remote location a mineable platinum zone would have to have at least fifteen grams per ton and be more than one meter thick.

Our trip shut down the operation. Keith wanted it done as quickly as possible in order to devote the remaining time for prospecting. By taking my canoe, we could get the job done in a long day and return to the Tree River camp late in the evening.

The Muskox Intrusion is a major structure in the Canadian Shield. There are three other nickel-copper-platinum bearing intrusions of similar size in North America: 1) the Sudbury Intrusion at Sudbury, Ontario where INCO has its principal mines; 2) the Stillwater Complex in Montana, where platinum is mined from a layer rich in chromite; and 3) the Duluth Complex, immediately south of the Minnesota/Ontario border, where uneconomic nickel-copper-platinum mineralization is present in a basal sulfide zone. All of these intrusions are layered, and all have basal zones of sulfide minerals and layers of iron, titanium, and chrome oxide minerals higher in the sequence of crystallization.

What is the origin of layered igneous intrusions of basaltic composition? All share four features: 1) the magma originated in the asthenosphere at a depth below 140 kilometers, and ascended

71

through fractures in the lithosphere; 2) the magma did not reach the surface; 3) the magma was emplaced in large reservoirs within the continental crust, probably at a minimum depth of six kilometers; and 4) it slowly cooled without being disturbed by faulting or the injection of additional magma while it was slowly cooling.

The exposed portion of the Muskox Intrusion is 120 kilometers long in a north-south direction and eleven kilometers wide at its greatest width. Its initial outcrop is a feeder dike (150 to 500 meters wide) that extends for sixty-five kilometers south of the intrusion. Internal layering does not begin until three kilometers north of the Coppermine River, which is 20 kilometers north of the Arctic Circle. Strong magnetic and gravity anomalies exist for an additional 230 kilometers north of the northernmost outcrops of Muskox rocks. Almost certainly, these anomalies are caused by the buried extension of the intrusion.

In cross-section, the Muskox Intrusion has the shape of a shallow funnel with its feeder dike at the bottom. The layered rocks of the intrusion are within the funnel. The funnel is a down-facing wedge of rock that dips fifteen degrees inward, and the sequence of layered rocks is 2.1 kilometers thick above the feeder dike. The intrusion has been tipped five degrees to the north, and an erosion surface obliquely slices through its total thickness, from its roof in the north to the feeder dike in the south. There is little doubt that 2.1 kilometers is the true vertical thickness of the exposed portion of the intrusion. The Muskox Intrusion has been intensively studied because of its simple structure, lack of deformation and metamorphism, and large areas of outcrop.

The layers of the Muskox Intrusion are within a shell of homogeneous rock (marginal zone) that was produced by the rapid cooling of the magma when it came into contact with the enclosing rocks of the crust. The rocks in the marginal zone preserve the original composition of the magma that filled the Muskox reservoir. It is accurate to say that the layering within the Muskox Intrusion has evolved within the rocks of the marginal zone, similar to the way that chicken embryos evolve within their shells. The composition of the shell is basalt, which indicates that all of the layers within the intrusion, taken as a whole, had the composition of basalt when they began their ascent from the asthenosphere.

The layers within the funnel are the result of slow cooling and

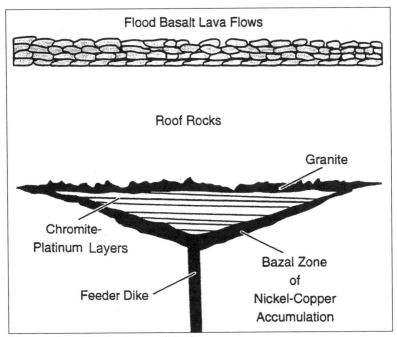

Flood Basalt Lava Flows

Roof Rocks

Granite

Chromite-
Platinum Layers

Feeder Dike

Bazal Zone
of
Nickel-Copper
Accumulation

Cross-section of Muskox Igneous Intrusion.

gravity differentiation. After the sulfide minerals crystallized and settled to the bottom of the reservoir, the heavier, iron-magnesium silicate minerals crystallized and concentrated in the lower layers; and the lighter calcium, potassium, aluminum, and sodium silicate minerals concentrated in the upper layers, with quartz concentrating in the uppermost layers. The layered series has forty-two mappable layers that can be as thin as three meters and as thick as 350 meters. Their outcrops form south facing arcs across the width of the intrusion. All layers tend to be thicker in the middle and thinner toward the margins where they merge into the homogeneous rock of the marginal zone. The position of layers within the sequence is highly predictable.

The silicate minerals containing the lighter elements of potassium, calcium, sodium, and aluminum were the last to crystallize. They are concentrated in an uppermost layers because gravity settling has removed most of the heavier silicate minerals and concentrated quartz. The uppermost layer has the composition of granite, and where it is in contact with roof rocks it is very rich in the mineral quartz. There is no quartz in the basal layers of the Muskox

Intrusion, but it is a common mineral in the uppermost granitic layers, where it constitutes 10 to 50 percent of the rock. Compared to all other layers in the intrusion, the granite layer has a highly variable thickness because, during the process of intrusion, large blocks of rocks with a high silica content were detached from the roof and settled into the granite magma where they melted and were wholly or partially dissolved. Quartz has the lowest specific gravity of any mineral in the Muskox Intrusion, therefore, the uppermost layer of granite, with its high quartz content, has the lowest specific gravity of any layer in the intrusion. By comparison, the basal layer of sulfide minerals has the highest specific gravity. Quartz and metallic sulfides are the two end member minerals of gravity differentiation.

The Muskox Intrusion is part of the Mackenzie dike swarm. The Mackenzie dike swarm is the largest of the thirteen dike swarms that have been identified on the Canadian Shield. All of the dikes in it have the same age as the Muskox Intrusion. Dike swarms are closely spaced, parallel diabase dikes having the same radiometric ages and similar contents of trace elements. Individual dikes within the swarm can vary in width from 1 to 500 meters and they can be as short as 100 meters or as long as 200 kilometers. They can also branch and bifurcate as upwelling magma find zones of weakness on its ascent through the crust.

In themselves, dike swarms are major igneous events, but they are always related to much larger geologic events. Later geological processes have usually erased these events because they occurred at distant localities, or there have been great movements in the earth's crust that have obliterated the related events. Some dike swarms, however, can be related to a geologic event that is still visible. Two of the best examples are: the dikes in the Bathurst Inlet rift zone, and the mid-continent rift zone centered in Lake Superior. Both of these events are discussed in the next chapter.

Mackenzie age dikes form a fan that is 1,500 kilometers long and approximately 1,000 kilometers wide. The apex of the fan is the south shore of Coronation Gulf, between the mouth of the Coppermine River and Bathurst Inlet, a distance of 300 kilometers. Dikes form a dense array south of Coronation Gulf. The dikes exposed in the Tree River drainway are part of that array. The density of dikes gradually diminishes toward the Great Slave Lake, 500 kilometers to the south.

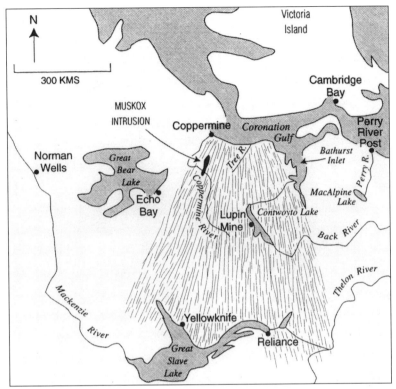

MacKenzie Dike Swarm

Outcrop areas of basalt lava flows are preserved immediately north and west of the Muskox Intrusion. They have the same age as the Mackenzie dikes and the Muskox Intrusion. It is clear that many Mackenzie dikes were feeders for Coppermine lava flows, in the same way that the dike exposed south of the Coppermine River was the feeder for the Muskox Intrusion. Basalt magma that reaches the surface through feeder dikes produces flood basalts. The extrusion of flood basalt is passive because it is very fluid, in contrast to lava extruded from volcanos that is often explosive. During episodes of extrusion, flows are rapidly deposited on top of one another. Where flood basalt flows are well exposed, some flows are more than fifty meters thick and cover tens of thousands of square kilometers. Flood basalts build plateau edifices, in short periods of geologic time (1 to 2,000,000 years), by depositing flows faster than they can be eroded.

The volume of magma that filled the Muskox reservoir is

indicative of the huge volume of lava that reached the surface and became the Coppermine flood basalts. The Coppermine series of flood basalts was extruded as a plateau on top of the roof rocks of the Muskox Intrusion. The plateau has an indicated thickness of three and a half kilometers. Many outpourings of flood basalt are well preserved at other localities. The preserved portion of the Coppermine flood basalt plateau is similar in every way to the Deccan flood basalt plateau in India that is of late Cretaceous age (65 million years), and to the Ethiopian plateau in East Africa (15 million years).

It is, therefore, a safe assumption that where Mackenzie dikes are closely spaced, as they are in the region south of Coronation Gulf, the basalt that ascended through them created a plateau with an area of perhaps a million square kilometers. Investigators familiar with the geology of the Muskox Intrusion, and related Coppermine flood basalts, have estimated that the flows were emplaced during 1 to 3 million years. How much basalt flowed to the surface through all of the dikes of the Mackenzie swarm? Only an estimate is possible, but it was probably in excess of 1 million cubic kilometers. Where diabase dikes are most closely spaced, as at the Tree River, they created a pile of basalt lava flows several kilometers thick.

The most important question is what event caused the fractures that tapped the zone of partially melted basalt in the asthenosphere and left as evidence the Mackenzie dike swarm, the Muskox Intrusion, and the Coppermine flood basalts. Several geologists who have mapped the geology of the western Arctic have speculated on that event. In the last chapter I speculated that there was a mountain building event due to continental collision, and the relaxation of the compression that built mountains opened fractures that tapped the asthenosphere. There is, however, no good evidence in the geologic record of North America for a continental collision 1.27 billion years ago. Evidence may exist on another continent that has rifted and drifted since the emplacement of the Muskox Intrusion and the Mackenzie swarm. As of now, however, there is no satisfactory answer for the emplacement of the Mackenzie dike swarm. All we can say with certainty is that the dike swarm is evidence of a major geologic event when a huge volume of basalt lava was extruded on the surface of the earth.

CHAPTER 10

The Ocean That Never Was

My first visit to Bathurst Inlet was on August 2nd. The helicopter traverse was almost due east from Tree River camp, inland from the south shore of Coronation Gulf. My principal target was the east shore of the inlet, where INCO's senior exploration geologist had found a gold showing two years previously. Fred and I took our lunch break at the northern end of Banks Peninsula, in the center of Bathurst Inlet. We stopped beside an outcrop of basalt lava flows and hoped we might find some native copper in it. The flows at Bathurst Inlet are known to contain occurrences of native copper of the same type that are found in flows exposed near the mouth of the Coppermine River. The Inuit had formerly searched for copper to make scraping knives and ornaments. INCO was not interested in native copper occurrences because a major deposit would have been found by the Inuit.

The outcrop surprised us. Between two lava flows was a bed of limestone one meter thick with some disseminated copper sulfides in it. I took a sample and spent thirty minutes trying to trace its continuity. It had no visible continuity, nor did it have a gossan, because it was exposed on a vertical rock face. We went on to the east shore to our prime prospecting area, but planned to visit the Banks Peninsula on our next traverse.

Bathurst Inlet is one of the deeper indentations in the northern edge of North America. From its mouth to its head, it is over 200 kilometers long, but over much of this distance it is a narrow, island-studded ribbon of the sea. Four large rivers empty into it: the James, Hood, Burnside, and Western. These rivers maintain a winter flow, despite their watersheds being wholly in the barren lands where precipitation averages 200 to 300 millimeters per year, and despite the land surface being underlain by permafrost. The water entering the rivers during the winter comes from streams that drain the largest lakes. It is pumped into the streams by thickening ice sheets. As ice sheets thicken and grow downward, they displace water that trickles into the streams and then into the river channels.

Bathurst Inlet is the most visible part of a continental rift zone that is over 400 kilometers long and about fifty kilometers wide. Rift zones are linear troughs of crustal weakness that are filled with younger rocks. The rocks that define the boundaries of the Bathurst Inlet rift zone are archean granites and gneisses (2.5 billion years old). They strongly resisted glacial abrasion so that they form boundary escarpments that frequently rise 200 meters from ocean level or from lowland plains. The boundaries of Bathurst Inlet are highly visible from the air. The diabase dikes, basalt flows, and sediments that fill the lowlands of the rift are 725 million years old. Gently folded basalt flows are preserved on Banks Peninsula and adjacent islands at the northern end of the Inlet. Elsewhere, they have been removed by erosion.

In the central and southern portions of Bathurst Inlet, the diabase dikes that delivered lava to the surface are the most visible evidence of the existence of a rift zone. In the Bathurst rift there are ten or twelve major diabase dikes, from 150 to 250 meters thick that have great north-south continuity. These dikes go north-south and overprint the Mackenzie dike swarm. Individual dikes vary, from vertical to sixty degrees and back again to vertical, and change thickness along their course. Where they become thin, their continuity can be observed as a string of low linear islands. The three or four thickest are continuously visible for 100 kilometers and, where they are thickest and are located in the middle of the inlet, they form linear peninsulas or linear islands that have continuous scarps on one or both sides. Alternatively, they are chains of sawtooth islands that are 150 to 200 meters high that protrude directly from tidewater, or are strings of sugar loaf mountains that jut from the lowland plain.

Continental rifts form when geothermal heat in the asthenosphere, at depths greater than 140 kilometers, is trapped in the zone of partial melting beneath continents. Geothermal heat is generated by the radioactive decay of the elements potassium, thorium, and uranium. This heat seeks a means of escape. The escape process begins when a vertical convection within the partially melted basalt (at the top of the asthenosphere) creates a dome that is fifty or more kilometers in diameter and five or more kilometers high. These domes are high enough to fracture the overlying lithosphere and the rigid continental rocks. Fractures go in three different directions (a triple junction) across the crowns of domes. When one or two of the

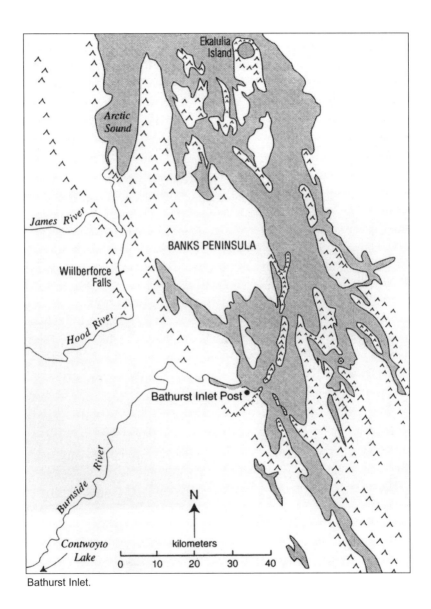

Bathurst Inlet.

fractures propagate, they lengthen and widen into rifts zones whose surface expressions are huge trenches in the crustal rocks that form continents. The length and widths of rifts are determined by the amount of geothermal heat that must escape. The fractures that define continental rift zones are the vents by which geothermal heat escapes in the form of flood basalt lava flows.

Continental rift zones usually propagate until they are long enough to vent enough heat to reestablish equilibrium between the amount of heat needing to escape and the number of fractures that allows heat to escape. The amount of heat that escapes is directly proportional to the amount of basalt that reaches the surface as flood basalt lava flows, or is emplaced as dikes, sills, or other intrusions in the upper fifteen kilometers of continental crust. As venting takes place, continental rifts subside in direct proportion to the amount of magma withdrawn from the zone of partial melting.

Several continental rift zones have been identified in Canada. The largest is the mid-continent rift shared with the United States. It centers on Lake Superior, but its southwestern arm extends 1,500 kilometers into central Kansas, and its southeastern arm extends 1,000 kilometers to the vicinity of Detroit, Michigan. For most of these distances, younger sediments cover the rift zone, but it has an unmistakable magnetic signature because the basalt flows and diabase dikes that fill the rift are highly magnetic. The Lake Superior portion of the mid-continent rift zone is filled with thirteen kilometers of lava flows, and sediments derived from the weathering of lava flows that built a high plateau of flood basalts 1 billion years ago.

Remnants of the basalt flows and related sediments that fill the mid-continent rift zone, (or were once part of the pile of flood basalts on top of the rift), are preserved in several places around the shore of Lake Superior. Extensive exposures are found along the north shore in Minnesota; in thick feeder dikes exposed south of Thunder Bay, Ontario; on Isle Royal and Michipicoten Island in the lake; and a few lakeshore outcrops north of Sault Ste. Marie, Ontario. The best exposures of basalt flows and sediments, however, are on the Keweenaw Peninsula of Michigan where native copper deposits were mined from the late 1840s to the mid-1960s. The native copper deposits of Michigan are similar to, but much larger than, the sporadic occurrences of native copper found in outcrops of basalt flows in the Coppermine River area, 300 kilometers west of Bathurst Inlet; and the equally sporadic occurrences of native copper in outcrops of basalt flows on the islands and peninsulas at the mouth of Bathurst Inlet.

Three other continental rift zones in Canada do not have exposed lava flows because they are covered with younger sediments, or because the flows have been eroded. These rift zones are: the lower St. Lawrence River valley from Montreal, Quebec to the Atlantic Ocean; the Ottawa River valley as far west as North Bay, Ontario; and the Timiskaming rift exposed at Cobalt, Ontario, but extending south to Lake Nipissing and north to James Bay. All of these rift zones were discovered after geologists recognized that the linear valleys and lakes of East Africa—Nyasa, Tanganyika, Kivu, Albert, Edwards, Rudolf, and Abaya—occupy continental rift zones. These lakes fill a series of enechelon rifts, 4,000 kilometers long, from central Mozambique to the Ethiopian plateau.

The East African rift zone and the Red Sea rift zone are the last activity of the breakup of the super continent of Pangea, that began about 200 million years ago when all of the earth's continents were joined. The biggest continental rift was the separation of North America from Europe. Separation began when a continental rift began widening at the rate of six or seven centimeters per year. The separation continues today, but at a rate of two or three centimeters per year. Beginning about 130 million years ago, another continental rift began to separate South America from Africa, and that separation also continues today at a rate of two to three centimeters per year.

The Atlantic Ocean came into existence when continental rift zones divided North America from Europe, and South America from Africa. The rifts began separately but joined and widened into the Atlantic Ocean. At the center of the Atlantic Ocean is the mid-Atlantic spreading center that extends for 15,000 kilometers from the Arctic Ocean to Antarctica. It is part of a global system of mid-ocean spreading centers that are continuously connected for over 60,000 kilometers, and which continuously produces new oceanic crust composed of basalt.

No oceanic crust is older than 200 million years, and 50 percent of all oceanic crust today is less than 65 million years old. All oceanic crust is recent (in geologic terms), compared to the oldest

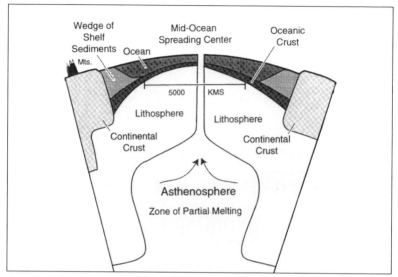

Structure of mid-ocean spreading centers *(not to scale)*.

rocks in continental crusts that are 3.9 billion years old. Mid-ocean spreading centers are difficult to investigate because they are hidden by four and a half kilometers of water. There is one exception. It is the island of Iceland where the mid-Atlantic spreading center is at the surface.

The continental rift zones that fragmented Pangea came into existence to vent an enormous amount of geothermal heat trapped under the super continent. When Pangea began to break up, separation occurred faster than basalt flows and sediments could fill the rift. When magma reached the ocean floor, it was extruded as pillow basalt flows. This continues today. Contemporary oceanographers, in deep dive vehicles, have photographed recently extruded pillow basalts on ocean floors. At mid-ocean spreading centers, magma that fails to reach the ocean floor solidifies in the fractures that were carrying it upward. These channels are preserved as closely spaced dikes that extend downward from the ocean floor for a kilometer or more. Each vertical dike is evidence of an episode of heat escape that created additional oceanic crust.

During the past 100 million years, the venting of geothermal heat has become concentrated in mid-ocean spreading centers where new oceanic crust is continuously created. Mid-ocean spreading centers are the preferred site for venting geothermal heat because oceanic crust is thin compared to continental crust. Oceanic crust averages seven kilometers and continental crust forty kilometers. When new oceanic crust began forming after the breakup of Pangea, it required less geothermal energy to induce fractures in it compared to continental crust. It also required less geothermal energy to propagate fractures. Therefore, almost all active rift zones today are at the centers of oceans where the crust is thinnest, and where the partially melted basaltic magma of the asthenosphere is closest to the surface.

All of our contemporary oceans have come into existence during the past 200 million years in the same way that the Atlantic Ocean came to existence, and all of the continents have moved into their present positions during this same time span. Yet, the venting of geothermal heat from mid-ocean spreading centers, and from all volcanos, annually contributes only 0.02 percent of the heat that reaches the surface of the earth. The other 99.98 percent comes from the sun. However, the geothermal energy originating in the outer mantle is powerful enough to move continents thousands of kilometers and create oceans where none previously existed.

What Fred and I saw at the mouth of Bathurst Inlet was a triple junction that had opened momentarily when a dome of partially melted basalt at the top of the asthenosphere had fractured the overlying lithosphere and continental crust. Two of the fractures widened and lengthened at a sufficiently fast rate to vent the accumulated heat. Judging by the thickness of the feeder dikes in Bathurst Inlet, there was an enormous outpouring of lava flows in a short period of geologic time (2 or 3 million years). This outpouring probably formed a plateau of flood basalts. A remnant of this plateau is preserved on Banks Peninsula, Ekalulia Island, and other islands in Bathurst Inlet. It indicates that the flood basalt plateau was at least two kilometers thick, but it could have been twice that thickness.

The evidence of this outpouring is that two arms of the triple junction, that preserve basalt, flowed onto a land surface. The most visible arm extends the length of Bathust Inlet. The second arm is preserved in basalt outcrops along the south shore of Coronation Gulf toward Coppermine River. The outpouring of flood basalts, in the Bathurst rift zone, in a short period of geologic time, vented enough geothermal heat to stop the process of ocean formation. Bathurst Inlet is an ocean that never was.

After spending three hours prospecting the eastern shore of the Inlet, we decided to have a late lunch on the shore of Ekalulia Island. Fred saw a small beach by the stream draining Ekalulia Lake where we could land and do some salt water fishing. Ekalulia Island is twenty-three kilometers long and up to eight kilometers wide. The northern half of the island is an oval lake about five kilometers in diameter with a shore that rises 150 to 200 meters from tidewater. Most of the island's outer face is a cliff, while the inner surface is a slope of bare rock that dips twenty to seventy degrees inward from the rim to the lake shore. The lake resembles the crater of a volcano that blew its top and was flooded by the sea, but this is not the explanation for its shape. The lake basin was formed by differential glacial erosion.

Before glaciation Ekalulia Island was a very large bulge in a diabase dike. The bulge formed where a vertical dike divided into a V as it approached the earth's surface. The division of the dike was probably caused by rising magma moving along a syncline in sediments. A V-shaped basin of sediments occupied the center of the V. Glaciation excavated a basin by gouging out the poorly consolidated sediments that filled the V. The glacier also undercut the outer surface of the bifurcated dike so that, when the glacier melted, the undercut rocks collapsed. Today, most of the island's shore is a cliff and the interior basin is filled with meltwater.

Our landing beach provided a surprise. There was a scattering of bleached whale bones above the high tide mark. They varied in size from a jawbone one meter long, to ribs and vertebrae that looked like they came from cattle. One of the bones was a badly chipped narwhal tusk. We lashed it on top of one float and the jawbone on the other. Then, we laid an assortment of vertebrae and rib bones beside them. We ate a hurried lunch and spent half-an-hour catching tommy cod, which we returned to the sea. The return traverse to base camp was uneventful.

CHAPTER 11

Visiting When Nobody is Home

Two days after visiting Ekalulia Island, I used the chopper to prospect a large part of Banks Peninsula. We landed beside outcrops of basalt lava flows looking for additional outcrops of the limestone containing copper mineralization. We found nothing. Our return flight to the Tree River camp paralleled the James River. The day had been fatiguing because the region was bathed in a great continental high. The air temperature was close to thirty degrees, which was intensified by sunlight heating the inside of our bubble cockpit. About ten kilometers west of Arctic Sound, we began looking for a flat rock outcrop where we could land and have a very late lunch. There were no interesting outcrops in sight until Fred spotted a white circle surrounded by small white blotches.

The site was an abandoned Inuit winter camp made by a family of caribou hunters from Bathurst Inlet. It was located at the southern end of a small lake where the uplands begin dropping into the James River valley, and where two converging bluffs created a funnel that channeled grazing caribou south to the river. We landed about twenty meters from the white spots and, while I was getting out, Fred turned off the engine and locked the controls. From the air the site looked sloppy, so I grabbed my hammer for turning over objects. Then, I tested the air.

Through the open door I said to Fred, "It's a bit hard on the nose."

"You mean there's a smell, even from here? If that's the case, I don't want to have lunch here."

"Okay, I'm with you on that, but let's look around first."

"Roger, Roger."

First we went to the white circle, and found that it was made of winter caribou hides, sewn together with caribou sinew, with the hair turned out. The circle was about two and a half meters in diameter, and the canvas tent had been no more than one and a half meters high at its peak. The hides had been placed against the sides of the tent in order to insulate it from the wind and cold, leaving only enough canvas at the peak to admit dim light. Another set of hides was used as a floor. The hides were now crumpled heaps with pools of water among the folds. The pools were breeding and feeding incubators for the larva of bluebottle flies, mosquitoes, and several other insects. The hides were so soft that the slightest jab with the hammer shredded them, and caused clumps of long white hair to come out in handfuls.

The ground immediately surrounding the tent looked like an abattoir of the type condemned by the earliest municipal reformers, and the interior of the tent was little better. Inside the tent was a tobacco can full of melted fat that rested on a flat rock that was covered with congealed fat. A layer of blue bottle flies covered both. Two caribou forelegs with attached hooves lay in one corner where they had been thrown after being partially stripped of their skin and gnawed. The fleshy parts were oozing grease and white maggots.

In another corner were some strips of hide and, under an adjacent fold, were several piles of excrement. Another fold concealed some sawed off antler tines, and nearby were half a dozen knee joints with dried flesh still clinging to them. They were shiny in the bright sunlight due to the oil oozing from them. Further around the tent ring was a pile of partially decayed wings, resting on a layer of split caribou leg bones from which the marrow had been scraped. This was mixed with several partially chewed ribs, and some thick strands of caribou sinew that had been scraped but not shredded. Apparently, they had been abandoned when the woman of the tent found she had enough thread to make her man a new pair of mittens.

Scattered over the floor were bits of bone, shreds of meat, and glistening patches of grease on sprigs of vegetation that had grown through the rotten hide floor. The tent had been pitched on top of fifteen centimeters of hard, wind-packed snow, and the snow beneath the hide floor had not melted during the winter. The underlying bushes and mosses had not been trampled flat by the people living over them because the snow was rigidly frozen when the tent was erected. The frozen environment must be kept in mind when the mode of Inuit living is reconstructed. Snow is a disinfectant. All the garbage that was spilled on the tent floor, and all the excrement in the corners of the tent were quickly frozen. Still, the living conditions appeared to be dark, crowded, and filthy, even taking into consideration the neutralizing qualities of the snow and cold.

The litter was astonishingly varied. Fred picked up a cast aluminum plate from an Irving parachute's release mechanism. It rested on a pilot's belly and the four harness straps fitted into it so they could be quickly released by pressing a central button. All of the springs and latches were gone and only the housing remained, but Fred recognized it because he had once owned one. He had lent it to a friend who had gone on a forest spraying contract in New Brunswick, but it had not been returned because the pilot crashed without having an opportunity to use it. Perhaps this casting came into Inuit hands under similar circumstances.

Inside the tent was a broken quart thermos jug, an empty can of cigarette tobacco, a split-open set of drycell radio batteries, and about three meters of yellow blasting wire that had been used as an aerial. Since we were 250 kilometers from the treeline, their cooking was probably done on a naphtha stove, as there were no wood ashes near the tent. Wood fuel, of a sort, was available from a nearby grove of scrub willows, but it would have had to be dug from under the snow. This was not done.

Near the entrance of the tent was a rusty brace with a dull bit still inserted. The flat wooden pressure pad was missing. Nearby, half-concealed in a fold of a decaying caribou hide was a small, rusty knife with its handle missing. Nearby, mixed in the refuse, I found a lashing divot (for ropes) that had been newly shaped from a caribou antler. The divot, if needed, would have been tied or nailed onto a tummytuk (Inuit sled) through the four small holes drilled into it. Also littering the floor were twenty-four black, curled-up negatives from a brownie camera.

Outside the tent, along the shore, was a pile of wooden slats from packing crates. They were used to make sled repairs and anchor traps in the snow. All traps have a chain with a ring attached to the loose end. A slat was inserted through the ring and then frozen into the snow. Lying in front of the slat pile was a rusting 22-rifle without a stock, and, lying on top of the wooden slats were, four rusty traps and five broken Thermos jugs. The traps were of two sizes: large ones for wolves or wolverines, and small ones for foxes. Nearby were half-a-dozen antlers with their tines sawed off, and an empty two gallon can of corn oil. There were also two piles of crushed, burned bones whose origin I cannot definitely explain. They were probably burned to get the ash that would be wetted and applied to sled runners (to get a smooth running surface). The Inuit will generally plan ahead for winter by storing a supply of mud to use on sled runners. Perhaps this family ran short and used ashes as a substitute.

The greatest litter around the tent was animal remains. Two meters from the tent circle was a skinned, dried fox carcass and two caribou heads with the antlers still attached. The tongues had been cut out and the eyes were no longer in their sockets, but otherwise they were complete. One had been placed on a rock where it had been exposed to the cold wind. It was dry and light. The other one lay in moist moss and was rapidly decaying. Within fifteen meters of the tent were seven other caribou heads, all of them in a state of rapid decay. Among the heads was a litter of knee joints, ribs, split leg bones, scraps of hide, hoofs, and many piles of excrement. Behind the tent were two piles of congealed, blackened entrails that had not been eaten by the Inuit or fed to the dogs.

The last ingredient to the pervading odor came from abandoned caribou hides. These were the white blotches we had seen while circling prior to landing. Winter hides are very thick and have hair that is seven to eight centimeters long. They are too heavy to be used for clothing. They are usually used as sleeping robes and tent insulation, or fed to dogs. There were more than a dozen discarded hides scattered around the tent site. Some were spread out and others were crumpled in the moss, but all were in a state of rapid decay. Those with flesh still clinging to them were covered with bluebottle flies, while those lacking flesh were smooth and shiny due to oil oozing to the surface. The long white hairs were wonderful homes for mosquitoes and black flies.

The next thing that caught our attention was a shallow, in-ground pit at the lakeshore that did not appear to be permanently frozen. Its bottom was completely covered with decayed wings with long feathers at their tips. It was not a garbage pit, but a pit where the family stored ptarmigan carcasses until they began to decay. After they had attained proper ripeness, they were exhumed and eaten.

Further down the shore was a line of short wooden stakes where the sled dogs had been tethered. All of the stakes had loose ends of ropes attached to them. When camp had been broken, the loops holding the dogs to the stakes had been cut instead of being slipped off. The ground in the area where the dogs had been tied was a solid mat of excrement, hair balls (from dogs being fed caribou hides), chewed bones, frazzled ropes, and feathers. The dogs were scavengers, and when the humans were done with bones the dogs got them, plus entrails and fox carcasses.

The man of the house was probably a fox trapper, or at least he did some trapping, because we found a pile of twenty unused traps on a rock about 100 meters behind the tent. The traps were a jumble of chains, springs and jaws, and it would have taken considerable time to unsnarl them. They still had spots of shiny newness among the general rust. The cardboard cartons they came in were lying nearby. At the base of the rock was a loop of wooden net bobbers with their original varnish unscratched, but there were no signs of nets. No fishing had been done because there were no fish bones among the mass of offal. The bobbers had come with a net, and the net had probably been used to wrap the family's possessions on their sled when they broke camp.

Apparently, caribou hunting was so good that the man of the house had not been forced to run an extended trap line in order to trade furs for food at the Hudson Bay post at Bathurst Inlet, nor did he have to spread nets under the ice to avoid hunger. Since the Canadian government provides free traps and nets for the Inuit, they could afford to leave them when they were not immediately needed and receive new ones the next year.

CHAPTER 12

The Great Esker

The Great Esker runs northwesterly for 100 kilometers, but because of zigzags, its true length is over 110 kilometers. It is the longest esker I have seen, and is probably one of the longest in the world. It is located about 800 kilometers northwest of the crown of the last continental glacier. It begins at the middle portion of the Hood River, and ends about ten kilometers south of the Tree River base camp. It is in the barren lands for its entire length and it is, therefore, easy to follow even when it becomes sand dunes.

Eskers are linear ridges of highly sorted sand, gravel, and boulders. They are formed by sediments being deposited in ice-lined channels (or tunnels), near the bottom of glaciers during the final stage of melting. Frequently, they have razor sharp crests that undulate along their length. The most spectacular portion of the Great Esker is where it is a single inverted V ridge that rises forty meters above the plain. In another segment, it is two parallel ridges ten to twenty-five meters high. In two other segments, it is a jumble of high gravel mounds and, in another segment, it is a plateau of sand dunes. The high gravel mounds and sand plateaus were formed where two or more crevasse channels intersected. An ice-walled lake formed at the intersection, and sediments carried in the ice-lined channels filled it. But, regardless of what structures make up the Great Esker, they are always aligned along a continuous trend so that there is no doubt of the esker's continuity.

The course of an esker bears little relation to bedrock topography. Its channel was controlled by the pattern of crevasses in the ice at the time of melting. Living continental glaciers, like Antarctica and Greenland, often move over very uneven land surfaces as they flow toward their margins. Uneven bedrock topography causes lateral stresses in the ice that are compensated for by ice ridges (pressure ridges) and crevasses. Ice ridges are often high, and crevasses very deep. Frequently they are in an enechelon pattern, where ice movement is over buried hills or around mountains. As long as a glacier is moving, crevasses continually open and close, but when a glacier stagnates, crevasses remain open and form drainage channels that are filled with meltwater seeking an outlet to the ocean.

As snowfall decreased and the climate warmed, the North American glacier continued moving toward its margins. Movement continued because the glacier's crown flattened, but flattening caused a decrease in the speed of advance. At this juncture, it began to recede because it was melting faster at its margins than the ice advanced. The outward movement of the ice stopped when the glacier's crown was reduced to a thickness of 500 meters. In the region of the Great Esker, this occurred about 9,200 years ago. The glacier became a stagnant ice field and melted in place, releasing a vast torrent of water to find its way to the ocean. For most of its length, the Great Esker river ran inside the glacier, well above the land surface. When it cut into the lower 150 meters of ice, meltwater began picking up sand, gravel, and boulders. These particles filled low spots in crevasse drainage channels. The channels may have rested on bedrock, but more likely there was a layer of ice beneath the bottom of the crevasse channel and the land surface.

The gravel and boulders that now form inverted V ridges were deposited as glacial fluvial sediments within the V-shaped crevasses. These sediments formed downward pointing wedges, with the stream flowing on top of the sediments. When the ice that supported the wedge of sediments melted, the wedges collapsed straight down to form high ridges. The streambed at the center of the channel became the crest of the esker, while the outer edges of the channel became the ground level margins of the esker. If the streambed consisted of boulders at the time of collapse, the outer surface of the esker is covered with boulders. If the streambed was gravel, the esker's surface is gravel. If the channel was filled with sand, the sand has been drifted into dunes by arctic winds.

The Great Esker was a glorious river in its ice canyon or tunnel. Its water was exactly at freezing point. During the peak period of summer melting, it was over thirty meters deep and 100 to 300 meters wide, and it moved boulders as if they were grains of sand. The highest portion of the present esker was the deepest portion of the crevasse channel, because the low spot had to be filled with sediments for the river to maintain its gradient.

Eskers are very temporary features, in the sense that their functional life of draining glacial meltwater is a few hundred years at most. The sand and gravel that were deposited in crevasse channels, however, are enduring geographical features. The Great Esker

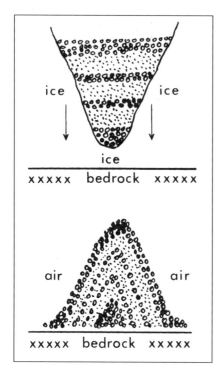

ended where it emptied into the ocean. The sand that it carried was deposited in a delta, but the rebound of land following the melting of the ice has elevated the delta above ocean level. The delta is now a sand plain that prevailing winds have heaped into the dunes, which were behind our tents at Tree River base camp.

We had flown over the Great Esker frequently while flying traverses to Bathurst Inlet. Fred and I grew fond of it, because on return traverses, when we saw its highest crest in its central portion, we knew we were not far from camp. It was a welcome landmark after three or four hours in the chopper, engulfed in the incessant roar of the engine, and bathed in the headache-producing glare of the sun coming through the bubble cockpit.

Keith decided to break Tree River camp on August 5th, and move 300 kilometers southeast to a lake a short distance north of the northern end of Beechey Lake. We laid out an indirect chopper traverse to the new camp, in order to make a refueling stop at a gas cache on an island at the southern end of Contwoyto Lake. We could also make an exploration traverse on the way. Our route took us 260 kilometers south, and then 110 kilometers east to Beechey Lake. At the beginning of the move, we used the Great Esker as a navigational aid since the initial leg of our traverse was nearly parallel to its course.

The Great Esker is a spectacular topographical landmark in its middle portion. It rests on granite that glacial abrasion has smoothed into an undulating plain. The highest segment, which is over six kilometers long, stands in accentuated height because the setting is so flat. Its highest section is forty meters above the plain. The high section is a single ridge with fifty degree slopes that rise in a single sweep. There are no sub-ridges or platforms to break its symmetry and continuity. The top and sides are covered with a layer of boulders. A few are nearly a meter in diameter, but most are about thirty centimeters. They are well rounded, but with irregular shapes.

We landed the chopper near the base of the esker, climbed to the top, and indulged in the arctic explorer's most sought after pleasure. We rolled boulders downhill. Unfortunately, there was no lake at the bottom where they could splash, nor were there any trees to be knocked over. But, we had a clear view of the fall, and were particularly gratified when a bouncing boulder dislodged another boulder and started it rolling, or a bounding boulder hit an embedded boulder and bounced high into the air, not hitting the ground until near the bottom of the slope. Fred and I spent half an hour rolling boulders, and then headed for the Contwoyto Lake weather station.

In order to fully appreciate the magnitude of the Great Esker, the whole process of continental glaciation must be understood. Why did land temperatures drop enough to form continental glaciers? Climatologists do not know, but there are at least four theories: 1) minor variations in the earth's orbit; 2) a decrease in the amount of carbon dioxide in the atmosphere; 3) the cyclical activity of sunspots that affects global weather patterns; or 4) a different pattern of currents in the north Atlantic Ocean that carried a much greater volume of tropical water into the Arctic Ocean.

The best explanation is the influx of tropical water into the Arctic Ocean by the Gulf Stream. The tropical water it carried prevented the surface of the Arctic Ocean from being sealed by a sheet of ice during the winter. If the Arctic Ocean remained open water during the winter, it would provide a continuous source of moisture for dry winds originating in Siberia. These winds would suck up moisture, and precipitate it as snow after crossing onto the cold landmass of North America. This cannot occur today because a permanent sheet of ice covers the Arctic Ocean. Whatever the reason, only a slight increase in the amount of tropical water entering the Arctic Ocean—enough to prevent the formation of an ice sheet— would cause the resumption of snow accumulation. The open water would provide sufficient moisture for snow to fall faster than it could melt during the summer, and the northern half of North America and Europe would again be covered by icecaps.

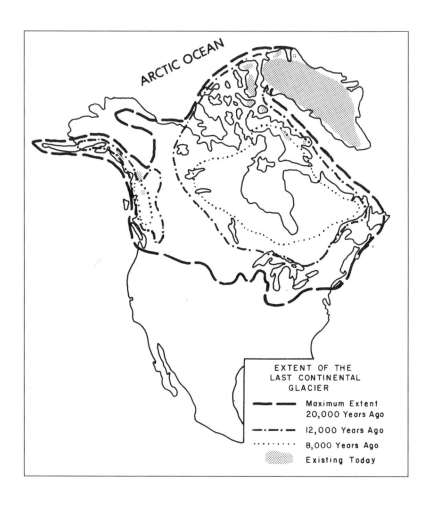

ARCTIC OCEAN

EXTENT OF THE
LAST CONTINENTAL
GLACIER

━━ ━━ Maximum Extent
20,000 Years Ago

━·━·━ 12,000 Years Ago

·········· 8,000 Years Ago

Existing Today

During the past 500,000 years, there have been four continental glaciers in North America and Europe. For continental glaciers to form, there must be more snow precipitation than now. Today, the annual precipitation in most of the Canadian Arctic is 300 millimeters or less. It is not enough for continental glacier to form. In fact, the Arctic would be a desert, without the prolonged freezing temperatures of winter that prevent evaporation. In particular, the ice sheets that cover lakes prevent lake water from being evaporated by the dry winds of winter.

When continental glaciers were growing, ice accumulated very quickly and within 25,000 to 30,000 years continental glaciers

94

reached their maximum extent. We know this because we know the rate of snow accumulation from cores drilled into the Greenland ice-cap. The slight melting of the surficial layer during the summer forms a distinctive layer, like annual rings in tree trunks. The melting of continental glaciers was also a sudden event, being completed in 25,000 years. We know this because the radio carbon dating of plant remains, preserved in bogs that formed on newly exposed land as the glacier receded, accurately measure the rate of recession. Both ice accumulation and melting were sudden events.

This interpretation is supported by the fact that the northernmost parts of the islands in the high Arctic, (islands west of Ellesmere and the northern coast of Greenland), are not glaciated, nor have they ever been glaciated. The northernmost fringe of North America and Greenland was never glaciated because the precipitation falling on it was mostly rain. Further inland, away from the warming effect of the open water of the Arctic Ocean, precipitation was in the form of snow.

In cross-section, the North American glacier looked like an overturned saucer. It was a continental size dome of ice that was thickest in the center but greatly thinned toward its margins. The ice moved toward its margins like pancake batter spreads in a frying pan, the thick center settling downward and forcing the semi-plastic ice to flow outward. At its maximum extent, the center of the North American glacier was over Hudson Bay, several hundred kilometers east of Churchill, Manitoba. It was probably 5,000 meters thick. Winter temperatures at its highest elevations were comparable to winter temperatures measured atop the Antarctic glacier at the South Pole. These temperatures are minus seventy degrees, and rise to only a few degrees above freezing during the three or four weeks of summer. Like the Greenland and Antarctic ice caps, snow constantly renewed its crown while the margins were continually melting. At both its southern margin, about 180 kilometers south of Chicago, and its northern margin, along the edge of the Arctic Ocean, the North American continental glacier was less than 150 meters thick.

As the glacier moved toward its margins, it scraped off soil and began abrading bedrock. The ice immediately above bedrock was packed with rock fragments. This debris acted like sandpaper on the earth's surface. In the long journey toward its margins, many of the rocks imprisoned within the glacier worked themselves up into the ice, or were there because they had been plucked from the tops of high hills. But, generally speaking, as one went higher in the ice, the amount of debris decreased. Only dust and micrometeorites were present 300 meters above the land surface. Near the margins of the glacier, where the rate of advance equaled the rate of melting, rock debris carried within the ice was deposited in irregular ridges composed of a semi-sorted mixture of sand, gravel, and boulders. These ridges are called terminal moraines, and they mark the margins of continental glaciation.

The continental glaciers of North America and Europe began melting about 30,000 years ago, and were completely gone by 5,500 years ago. This is a very large and abrupt climate change, but it was not a singular event. The geologic record documents repeated continental glaciations and repeated melting of continental glaciers in North America and Europe during the past 500,000 years. What were the most likely causes for the rapid onset of continental glaciation and the equally rapid melting of continental glaciers? Almost certainly, continental glaciation in North America and Europe was caused by an influx of tropical water into the North Atlantic and Arctic Oceans that prevented the formation of winter ice sheets. As of now, there is no good explanation why currents carried larger volumes of tropical water into the North Atlantic and Arctic Oceans than at present.

There is, however, good evidence for a trigger mechanism that initiated the rapid melting of the continental glaciers of North America and Europe. The triggering mechanism was a vast release of methane (natural gas) into the atmosphere. It came from methane hydrate. Methane hydrate is ice that forms in marine sediments at depths below 500 meters where the temperature is minus three or four degrees. Within its crystal structure, this ice can sequester several thousand times its volume of methane. Within the past thirty years, oil exploration on continental shelves have found very large numbers of methane seeps that are capped by huge deposits of methane hydrates. Pockmarks and blowouts formed by the eruptive releases of gas are common structures in sediments forming continental shelves. Petroleum geologists estimate that the methane trapped in hydrates contain more methane than any other single gas resource.

How was methane contained in hydrates released into the

The Shoestring Lake base camp on sand flat with widely spaced giant trees growing on it. Sheet ice still covers the lake.

Helicopter prospecting near the treeline.

Above: A gossan discovered from the air near the treeline.

Left: Sampling the gossan.

A Beaver airplane, with canoe tied to floats, ready for takeoff to establish a fly camp.

A Beaver airplane leaving a ground prospecting team in a fly camp.

A cliff formed by a diabase dike, Bathurst Inlet. Note, raised beach lines at the head of the bay still in process of rebounding.

A helicopter landing to sample a quartz vein, Bathurst Inlet. Note the helicopter for scale.

The abandoned winter camp of a caribou-hunting Inuit family. (*See chapter 11*)

Above: The Benchmark Lake base camp on the ground.

Right: The Benchmark Lake base camp from the air.

The Benchmark Lake base camp with helicopter and Beaver ready for their daily flights.

Above and below: A grove of black spruce trees, growing far out in the barren lands, on the micro-environment of a sand esker that winds have shaped into dunes.

The Contwoyto Lake base camp established to evaluate a gold prospect.

Above and right:
A Beaver aircraft supplying Contwoyto Lake base camp with aviation gas for the helicopter (red barrels).

The discovery outcrop on top of the hill overlooking Contwoyto Lake.

Top inset: Getting ready to sample the discovery outcrop.

Bottom inset: The author sampling the discovery outcrop.

Using a portable percussion drill to make holes in rock pavement, in order to blast a trench to sample the gold content of the rock.

Setting up the diamond drill to test the gold-bearing rock where the magnetometer survey indicated that it was widest.

A diamond drill drilling a hole at a 35 degree angle.

A driller emptying the core barrel after retrieving it from the end of the drill rods.

The home office geologist examining the drill core.

The home office geologist making a plane table map of the mineralized area.

The first two completed buildings of the permanent base camp erected during the final days of the 1961 field season at Contwoyto Lake.

The Coppermine River at the treeline where the feeder dike of the Muskox Intrusion crosses the river.

Barrenlands just before breakup.

atmosphere? It occurred after ocean levels fell by 120 meters due to the vast withdrawal of marine water locked into glacial ice. As ocean levels fell, the pressure that preserved hydrates lessened. The hydrates melted. Suddenly, worldwide, but especially in the northern hemisphere, there were incredibly huge bursts of methane entering the atmosphere from young, gas rich sediments that form the continental shelves of eastern North America, the North Sea, Norwegian coast, and the arctic coast of Siberia in Russia.

Ice cores drilled into the Greenland and Antarctic icecaps record polar weather conditions for 100,000 years. Gas bubbles trapped in this ice confirms high concentrations of atmospheric methane at the time of rapid melting. Under current atmospheric conditions methane takes about twelve years to oxidize to carbon dioxide. Methane is twenty times more efficient than carbon dioxide as a greenhouse gas, and the end product of oxidation is more carbon dioxide in the atmosphere – which is the usual reason cited by environmental activists for global warming during the past 100 years.

Computer modeling of paleoclimates indicates that bursts of methane entering the atmosphere persist longer than twelve years if the concentration is ten to twelve times greater than current levels. It can persist for 100 years. The combination of methane, more carbon dioxide from the oxidation of methane, and more water vapor in the atmosphere from greater evaporation due to warmer temperatures, created a greenhouse effect that was magnitudes greater than current temperature fluctuations. In north polar regions temperatures increased 7 to 10 degrees in less than 50 years. In terms of geologic time, glacial melting was an instantaneous event, and it took place when man was just another animal in the ecosystem. This greenhouse effect persisted until marine organisms captured much of the additional carbon dioxide in their carbonate shells after it was dissolved in marine water. Marine organisms have reduced the greenhouse effect of methane bursts to current levels. These shells are now part of shelf and ocean bottom sediments.

At the same time, two other events took place. 1) The rise in ocean levels reduced the volume of currents carrying tropical water into the North Atlantic and Arctic Oceans. 2) The rapid release of zero degree fresh water from melting glaciers formed a layer of lighter water on top of heavier marine water. The melt water froze much more easily than heavier salt water and an ice sheet formed on the Arctic Ocean; and the winter ice sheet on the North Atlantic extended further south. These ice sheets ended a source of snow sufficient to reactivate continental glaciation – at least for the present.

During the great melt, sea levels rose about four meters per 1,000 years. The melting of the North American and European glaciers, within 25,000 years, is the real global warming. It is quantifiably different from the minor temperature fluctuations of the current weather cycle.

It is hard for us to realize the vast volume of water that was released when the continental glaciers melted in an instant of geologic time. The liquefaction of the 600 or 700 million cubic kilometers of ice, in the European and North American glaciers, caused the ocean level to rise about 120 meters. And, if the existing icecaps of Greenland and Antarctica were to melt, ocean level would probably rise an additional seven or eight meters. The rise in ocean levels had unpredictable effects. It altered the shapes of landmasses, radically changed climates, effected the distribution of human populations, and contributed to the abrupt extinction of a large number of large mammals in North America and Europe. Some of the large mammals that were candidates for extinction by our hunter-gatherer ancestors, but survived, are horses, camels, cattle, llama, and alpaca. They survived because they were domesticated. Mammoths, mastodons, hairy rhinoceroses, giant ground sloths, and giant beavers were not domesticated and were hunted to extinction.

During the period of maximum glaciation, the Mediterranean Sea was also about 120 meters below its present level. Streams draining into the Mediterranean incised themselves into the continental shelves. For example, the Nile River incised a deep canyon into its delta, and the Bosporus strait, that now links the Sea of Marmara to the Black Sea basin, was a valley that drained into the Black Sea basin. The western end of this valley was eroded beneath the level of the pre-glacial ocean. By 7,500 years ago (5500 b.c.), at the end of glacial melting, the Mediterranean Sea rose high enough to overflow the Bosporus dam; downcut the Bosporus valley into a canyon; and pour an avalanche of marine water into the Black Sea basin, because the Black Sea basin was somewhere between 105 and 150 meters below the current ocean level. The Black Sea basin was below current ocean level during the period of maximum glaciation, because evaporation was greater than the inflow of rivers draining the glacial north.

Humans living around the perimeter of the fresh water lake that occupied the Black Sea basin enjoyed a warm climate, because the thicker layer of atmosphere converted sunlight into more heat than at ocean level. The lake was an oasis from the cold desiccating winds blowing from the glacial north. After the Bosporus dam was breached, the flood of marine water raised the water level about thirty centimeters per day, and converted the Black Sea lake into the Black Sea. In an instant, humans were forced to flee to higher ground.

It is highly probable that the people living around the periphery of the Black Sea lake practiced some form of village-centered agriculture. It is equally probable that Sumerians were among those cultivators, and that they practiced some form of irrigation. The sudden catastrophic flood forced them to flee. Migration took them to the sparsely populated Tigris-Euphrates valley, where the archaeological record indicates that their earliest villages were settled about the time of the Black Sea flood. There they applied their irrigation skills to the arid soils of the valley. The food surplus they produced was the foundation for the earliest urban civilization. Urban culture and commerce created a demand for literacy, and the Sumerians were among the earliest cultures to invent a script (cuneiform).

The flood that filled the Black Sea basin was a defining event for Sumerian culture, and its memory was preserved in the epic poem Gilgamesh, which was recited as long as the Sumerians existed as a distinct people. The purpose of the Gilgamesh epic was to explain why the Sumerians were forced to migrate, and why they had a priority claim to the land where they settled. Soon after the invention of cuneiform, the oral account of the flood was written and a true copy was stored in the king's archive. Abraham, one of the Jewish patriarchs, came from the Sumerian city of Ur (Genesis, Chapters 9-11). After leaving Ur, he joined a band of migrants who were seeking to establish themselves in a new land (in order to practice agriculture), and he brought the story of Noah surviving the flood with him (Genesis, Chapters 6-8). The Genesis account is a variant of the Gilgamesh epic and, when linked with God's promise to Moses (arable land to righteous survivors), served the same purpose as it served the Sumerians.

Other geographic changes followed the melting of continental glaciers. During the ice age, northern Africa was a well-watered savanna that supported as many animals per square kilometer as the Serengeti Plain does today. Today that savanna is the Sahara Desert. The rise in ocean levels also flooded the Indonesian plain so that it became the Java Sea and, in so doing, isolated elephant, rhinoceros, and tiger populations on the islands of Sumatra and Java. Likewise, the flooding of the Bering plain, between Siberia and Alaska, severed the land bridge that had allowed mammoths, bisons, wapiti (American elk), moose, caribou, big horn sheep, mountain goats, grizzly bears, lions, panthers, wolverines, and humans to migrate to North America. The same bridge had also allowed horses and camels to migrate to Eurasia.

Eurasian mega-fauna entered the interior of North America during warm inter-glacial periods, during the Pleistocene era (1.7 million years to 10,000 years ago). The first humans did not enter the interior of North America until 10,000 or 12,000 years ago when they migrated down the coast of Alaska and British Columbia, or down an intermittently ice-free corridor between the mountain glaciers of Alaska and British Columbia, and the western edge of the continental glacier. When sustained melting opened a permanent passageway, more bands of hunter-gatherers, speaking different languages, joined the earlier arrivals. They found a Garden of Eden populated by animals that had no fear of humans. Within 2,000 years of the arrival of human predators, 70 percent of the indigenous mega-fauna in North America was extinct. The arrival of human predators in South America, Australia, and New Zealand had a similar impact on indigenous mega-fauna. The large mammals that replaced extinct indigenous species were those that preceded or accompanied the human migration from Eurasia. These animals feared humans and survived.

As humans migrated south, they had a feast as long as there were indigenous large mammals to be killed and eaten. The extinction of these animals ended easy subsistence. The necessary response was inventing agriculture in order to produce adequate food to feed bands of hunter-gatherers, who increasingly competed for space after they had exterminated their accustomed source of food.

In anticipation of us stopping at the Contwoyto Lake weather station, Keith had our plane bring the weathermen's mail from

Yellowknife. Unless there was a special delivery, like ourselves, their mail arrived once a month with food supplies. The mail we brought consisted of ten letters for four men, and a drive shaft for an outboard motor.

The weather station is one of the few places between the DEW Line and Yellowknife where weather information was gathered in 1960. At the time, it was operated by a charter airline that supplied DEW Line sites. Their big planes flew north from the paved strip at Yellowknife to Cambridge Bay. From there, the supplies were distributed to outlying sites by smaller aircraft. The Department of Transport paid most of the station's expenses. The station's normal complement was three men, but a new man was learning the routine before an old hand left to visit his native Wales. He was due for a vacation, having spent three years at the station with only two weeks' vacation, every three months.

Our visit was as much a holiday for them as for us. We had been in the barrens six weeks without seeing a new face, so we gladly spent an hour drinking numerous cups of coffee and listening to their tales. They acted as a clearing house for the local Inuit population, half a social center and half an emergency station where the Inuit could summon aid.

The Contwoyto Lake band of Inuit were inland dwellers who live on the caribou herd and resort to fish and other food sources only when the herd fails. The herd is a fickle source of food at best. There are about six bands who live in the 150,000 square kilometers surrounding Contwoyto Lake and, during the year at least one of these bands will experience an emergency of some sort.

In February 1959, when daytime temperatures were minus forty-five degrees, a sled arrived with a request for the Mounties to help search for a sixteen year old boy who had been missing four days with his dog team. He had been out looking for arctic willow twigs in the sheltered valleys, when a blizzard caught him. When the storm abated, he was not found where he should have been. The RCMP flew in and searched, but found nothing. His sled was found three weeks later, and the Mounties had to return and hold an inquest. When the lad had seen the blizzard coming, he had started for camp, but the weather had overtaken him. He went blindly forward hoping to locate the camp, but he panicked and continued until his three youngest dogs dropped from exhaustion. He continued pushing the sled, while the lead dog pulled until it too was exhausted. The boy then huddled on top of the sled, drew the dog next to him for warmth, and pulled a caribou robe over both of them. The boy died first. The dog ate the caribou skin, and then died next to the boy.

Not all emergencies are this grim. The federal government is a great protector of the northern indigenous population. Younger members of families attend boarding school at Yellowknife for a couple of years. If older members contract tuberculosis, as they often do in their winter tents, they are sent to a hospital to be cured. If emergency medical attention is needed, a doctor is flown in. Because the Inuit are conscious of the help available, it is sometimes abused.

One message, written by a band's scribe, and transmitted from the Contwoyto Lake station read:

Dear Fokes:

Akoatuk him near dead in suffer in kinne. Him pretty near sick with troud and hine quarter not work. Tell doc in Yellowknife that him bad man an very sick an him come soon an take from bed when him cant move.

~ Mona Makoatuk

A doctor was flown in, but when he arrived the patient, who was supposed to have an acute throat and kidney ailment, was out hunting. When caribou became plentiful, he changed his mind about wanting a hospital vacation!

In an effort to reduce or end emergency food flights during the winter, the government decided to teach the Contwoyto bands to use nets, so they could feed fish to their dogs, instead of caribou meat, and dry some fish as an emergency food supply when the caribou herd migrated by a different route. The Department of Northern Affairs flew in a load of boards and a carpenter, who took two weeks to build two boats and teach them how to use them. Then, a government biologist taught them to use the nets from the boat in the summer, and how to spread them under the ice in the winter. At the same time, he made a study of the fish stock in Contwoyto Lake and found it large enough to be of commercial importance, as well as large enough to support the Inuit of the area. Someday trout from Contwoyto Lake may be served on Toronto's tables, just as trout from the Great Slave Lake are now marketed there. During the summer, the Inuit occasionally use their boats to catch some fish and, less occasionally, split and dry a few. But, when caribou are plentiful (as they were when we visited the weather station), the human and dog population live on them. In the winter, they will hunt until there is only one strong man and one strong dog team left in the band. Then, he will go to the Contwoyto weather station and summon a planeload of emergency food.

CHAPTER 13

Rainbow Lakes

Our new base camp was located 300 kilometers southeast of the Tree River camp, on an unnamed lake six kilometers north of Beechey Lake. It was four kilometers long and a kilometer wide. We called it Benchmark Lake because there was a geodetic survey marker on a hill near camp. Our target area was the east shore of Bathurst Inlet along the southern half of its length. We had already prospected the northern half from the Tree River camp, but we had not prospected the southern portions.

The helicopter would fly a fan-shaped pattern of prospecting traverses from Benchmark Lake. Both the morning and afternoon traverses would be about 150 kilometers long, and we would narrow the distance between flight lines when we prospected the eastern shore of Bathurst Inlet. This plan called for accurate navigation in order to prevent prospecting the same ground twice. Only after we had prospected the eastern shore of Bathurst Inlet, and the land east of Contwoyto Lake and north to the Burnside River, would we be free to make longer prospecting traverses to MacAlpine Lake and Perry River. It was not until August 16th that we could make a long traverse to the east. Keith and I had shared the flights to the east shore of Bathurst Inlet, but I made the flight to the rainbow lakes.

The rainbow lakes are 450 kilometers north of the treeline, in one of the most barren, remote, and uninhabited portions of North America. Inuit and caribou shun it because plants higher than twenty-five centimeters cannot grow there. Migratory birds, however, are vastly abundant during the summer, and a few muskox survive because of the region's extreme isolation. Fred and I discovered the rainbow lakes at the end of our first traverse. The lakes are located just north of the height of land separating the watershed, of the Ellice and Perry Rivers, from the Back River to the south. More precisely, they are found only on the MacAlpine plain, a drained lakebed that had been the western bay of a much larger glacial MacAlpine Lake.

Glacial MacAlpine Lake had a momentary existence during the great melt. At its maximum extent, it covered more than 6,400 square kilometers, but the present MacAlpine Lake occupies about 400 square kilometers of the deepest portions of glacial lake MacAlpine. The only outcrops on the MacAlpine plain are the tops of hills that were not buried by glacial-fluvial sediments. They barely protrude above the plain. The only other features that break the plain's flatness are low curving sand ridges that were beaches, and some shallow ravines adjacent to a no-name river that is slightly incised in the MacAlpine plain.

Instead of being a featureless plain, as are most drained lakebeds, the MacAlpine plain is studded with lakes. They are of all sizes. Invariably, they have round or oval shapes, and they are closely spaced so that they frequently touch each other. The lakes must be seen from the air, under specific weather conditions, for their colors to be seen. On a normal summer day, their waters are clear and they reflect the translucent blue of the sky, like every other arctic lake. But, if they are seen in July or August, on a cloudy day, when a twenty-five to thirty kilometers per hour wind is blowing, the lakes have colors that are independent of the reflected blue sky.

The uniformly shallow depths of these lakes allow winds to generate convections, which agitate bottom sediments causing variously colored rock flour, clay particles, and fine sand to go into suspension. The water in all of the rounded lakes is opaque, and their colors are pastel. Coloration depends on the color of the bottom sediments of each lake. Pink is the most common color, followed by greenish, cream, yellowish, and a barely perceptible shade of brown. The pinks and greens vary, from pale to a tinge so vague that it blends into cream or buff. Variously tinted lakes are found side by side, greenish next to pink, next to cream, next to buff.

We must explain the very unusual occurrence of many round lakes on a drained lakebed, and their strikingly similar appearances. Secondly, we must explain why the waters of these lakes are multicolored. These explanations require that we trace the immediate post-glacial history of MacAlpine Lake, and the basin it occupies.

Eight or nine thousand years ago, the retreating glacier exposed the MacAlpine basin. The basin had existed before continental glaciation, but the northward moving ice enlarged and deepened it. When it was re-exposed, it became a lake eighty kilometers wide and eighty kilometers long. After exposure, the glacier paused in its retreat for 200 or 300 years. The glacier's stationary front was along the southern rim of the MacAlpine basin and, when glacial MacAlpine Lake was in full flood, its waters were in direct contact with the glacial front. The glacial front was a serrated ice cliff 60 to 100 meters high, and icebergs calved from it and floated into the lake. During the summer, on average, about two meters of ice melted from the glacier's surface, from an area of 100 by 600 kilometers, and this meltwater drained into glacial MacAlpine Lake. The annual volume of water unlocked from this area was about 120 cubic kilometers, with most of it flowing into the lake during July and August. During these months, about two cubic kilometers of meltwater entered the lake every day.

During maximum melting, glacial MacAlpine Lake was always at flood stage. It was probably more than seventy-five meters deep, because the lake received water faster than the Perry River drainway could carry it to the Arctic Ocean. During July and August, the Perry River drainway was probably one kilometer wide, more than ten meters deep, and flowing at twenty-five kilometers per hour. Its erosive power was very great, so that the basin's northern outlet was constantly being lowered; first by eroding a trench through ground moraines, and then cutting into bedrock. This is the reason that the channel of the Perry River is vastly larger than the miniscule flow of water that it presently carries to the Arctic Ocean.

During the 200 or 300 years that the glacier remained stationary, the MacAlpine basin received about ninety cubic kilometers of glacial-fluvial sediments. Using a different measure of comparison, within these years, the MacAlpine basin probably received an average thickness of fifteen meters of sediments. The southern portion, which abutted against the glacier, was filled to overflowing, receiving more than ninety meters. The central portion was filled to an intermediate thickness, while the northern portion received comparatively little infill. Then, the lake drained.

The origin of the rainbow lakes begins with how glacial-fluvial sediments were carried into MacAlpine Lake. Glacial-fluvial sediments are rock particles that are transported by rivers issuing from the fronts of melting glaciers. They are often transported in ice-lined rivers that become eskers after the glacier has melted. The basal material of the MacAlpine basin is tillite that the melting glacier haphazardly dumped on bedrock. Within the tillite were many blocks of ice. Glacial-fluvial sediments then covered the tillite and blocks of ice.

During maximum summer melting, many of the meltwater rivers entering glacial MacAlpine Lake carried huge volumes of water, which were heavily laden with all sizes of rock particles gathered from the bottom layer of the glacier. When the meltwater rivers entered glacial MacAlpine Lake, they immediately dumped the boulders and cobbles they were carrying. They now form a high ridge that originated as a series of overlapping alluvial fans, deposited against the stationary ice front. All of their boulders are washed clean of smaller particles. It is geologically strange to see a pile of boulders, many over one meter in diameter, forming a high ridge, instead of being the basal layer buried by finer grained sediment.

After the boulders were dumped, coarse gravel, fine gravel, and coarse sand were deposited further out in the lake as the meltwater rivers lost velocity. In the center of the lake, medium-grained sand settled to the bottom, and in the northern part, fine-grained sand and rock flour were deposited in size-sorted layers. The deposit of the finest particles, however, was not uniform. The bottom currents that carried the finest particles had to flow around stranded icebergs and, as they lost velocity, these sediments were deposited, as lenses and tongues of fine brown sand, next to lenses and tongues of slightly oxidized silt. Both were overlaid by sheets of red, green, or white rock flour that was deposited during the winter when the sheet of winter ice stilled the water.

In the still water beneath the ice sheet, the finest particles of rock flour settled to the bottom. In spite of the cold temperature of the water, the rock flour that was deposited during the summer underwent slight oxidation, because the water was highly oxygenated, and because of the mixing effect of currents. After oxidization, these sediments were various shades of pink. Rock flour particles that were deposited from still water beneath the ice sheet, did not oxidize and retained their original greenish color.

The meltwater rivers, that carried boulders, also transported very large blocks of ice. They joined the icebergs that calved from the stationary ice front, and floated into the central and northern portions of the lake. During autumn and winter, meltwater ceased entering the lake. The lake level dropped and they were stranded. After being stranded, fine sand and rock flour accumulated around their bases.

Stranded icebergs could remain unmelted for many years, because the glacial water entering MacAlpine Lake had no extra heat. Meltwater entered the lake at exactly zero degrees, after flowing through ice-lined channels. It was already as cold as it could get without freezing. In autumn, after melting ceased, water continued to drain from the lake. The lake's level dropped ten or twenty meters, and most of the surface of stranded icebergs was exposed to frigid winter air. No melting occurred in the frigid winter air, nor in the zero degree water below the ice sheet.

It was not until glacial MacAlpine Lake ceased receiving meltwater that its water could accumulate enough heat, during the summer, to melt the stranded icebergs and the permafrosted tillite beneath the lake. The stranded icebergs could not melt, nor could the permafrosted tillite beneath the lake thaw, until the lake water warmed to four degrees. In the previous description of breakup, we have seen that winter ice sheets are very efficient insulators against refrigerating winds. They preserve the temperature of freshwater at four degrees, which makes lakes into reservoirs of heat. The heat put into MacAlpine Lake, by summer sunshine, melted the icebergs, and then thawed the permafrosted tillite.

After stranded icebergs melted, they left depressions in the bottom sediments. These depressions were partially filled by the slumping of the fine sand, silt, and rock flour that had accumulated around their bases. Slumping rounded the contours of the depressions. Stranded icebergs were not the only means of forming depressions in the bottom sediments. Earlier in this chapter, I described the imprisonment of blocks of ice in ground moraines. Permafrost preserved them, because there is no heat in the water of glacial MacAlpine Lake to melt the permafrost.

After glacial MacAlpine Lake ceased receiving meltwater, the permafrosted tillite thawed and the enclosed ice blocks melted. The overlying tillite and glacial-fluvial sediments collapsed and crater-like depressions were formed in the lake bottom. When glacial MacAlpine Lake drained, these depressions remained as pothole lakes.

The rounded depressions in the lake bottom were preserved because, in a geologic instant, glacial MacAlpine Lake stopped receiving meltwater and accompanying glacial-fluvial sediments. The volume of water entering MacAlpine Lake was reduced from a flood to a trickle, because the retreating glacier uncovered the Back River channel to the south, at a lower elevation.

After the MacAlpine basin stopped receiving meltwater, its area rapidly contracted. Glacial geologists have estimated that 400 or 500 years after glacial MacAlpine Lake stopped receiving meltwater, the lake was less than 20 percent of its original size. Today, MacAlpine Lake is only about 6 percent of the size of glacial MacAlpine Lake when it was receiving glacial meltwater.

What caused glacial MacAlpine Lake to drain? We have already mentioned four contributing causes: 1) the vast influx of meltwater, during the height of summer runoff inflated the size of the lake, and the autumn outflow reduced its size; 2) the enormous summer outflow had great erosive power that lowered the Perry River drainway; 3) recession of the glacier uncovered the lower Back River drainway; and 4) sediments filled the basin. None of these reasons, however, either alone or collectively, can account for the rapid contraction of the lake. There is a larger explanation. That explanation is earth rebound (isostasy).

When the North American continental glacier was at its maximum extent, there were 3,000 to 4,000 meters of ice directly over the MacAlpine basin. This load caused a flexure in the crust that was absorbed by the plastic asthenosphere, at a depth of 140 kilometers. After melting, the disappearance of an enormous weight of ice allowed the crust to rebound. The rebound occurred very quickly, most of it taking place 400 or 500 years after the ice melted. In the region of the MacAlpine basin, the rebound was more than 200 meters.

The amount of rebound is easily measured. When the glacier was in its final stages of melting, much of the northern rim of North America was underwater because the weight of ice had depressed the land below ocean level. Secondly, the huge volume of glacial meltwater that flooded into the oceans caused the ocean level to rise. When the land was most depressed, and the ocean level had risen fifty or more meters, marine water penetrated inland for considerable distances. When the ocean level stabilized, a sand beach was formed. These beaches are lines of equal elevation. The shells of marine mollusks, that lived in these beach sands, have been preserved. The beachline containing marine shells that is furthest inland marks the maximum extent of marine flooding. During the final stages of the melting of the North American glacier, rebound accelerated and marine flooding receded to the current coastline.

In the region of MacAlpine Lake, the furthest marine penetration is 130 kilometers inland from the present coastline. An elevated beachline marks this event. It is now 195 meters above the present ocean level. A few kilometers south of this marine beachline is another elevated beachline, which marks the northern shoreline of glacial MacAlpine Lake. It contains shells of freshwater snails. It is twelve meters higher than the marine beachline. Further south, this same beachline containing the shells of freshwater snails is twenty meters higher than the marine beachline. Thus, the southern part of the MacAlpine basin has rebounded eight meters more than the northern part. The effect of differential rebound was to tip the basin of glacial MacAlpine Lake northward, and accelerate its drainage.

Lake water spilled down the Perry River drainway. A combination of differential rebound and lowering the outlet by erosion, caused the central and southern portions of glacial MacAlpine Lake to drain almost as soon as meltwater ceased entering the lake. The northern portion became a shallow bay, but it too eventually drained. The round depressions in the drained lake bottom were preserved, because most of glacial MacAlpine Lake drained before additional sediments could fill them.

The slight difference in the color of the sediments, in the lakes, is related to the time of year they were deposited. The reddish and pinkish beds were deposited during the summer, when the oxygen content of the water was highest; the milky and greenish lenses were deposited during the autumn and winter; and the brown lenses were due to the fact that it was the color of unaltered, fine-grained sand. On windy summer days, these sediments go into suspension and impart various shades of pastel colors to the lakes. Occasionally, a normal, clear water lake is found on the MacAlpine plain. This is due to two factors: the lake either occupies a small basin on top of an outcrop of bedrock, or the lake bottom is coarse sand that does not go into suspension.

There is a single exception to the pastel coloration of the lakes: a lake with black, opaque water. We examined it because the shoreline was composed of black rubble. The rock was a peridotite intrusion that I immediately identified from the air. Typically, peridotite is a rock that disintegrates to black rubble when there is freeze-thaw weathering—the usual form of weathering in the Arctic. It is also a rock that is occasionally associated with nickel mineralization.

Black rubble was one of the principal features we looked for while prospecting from the air.

We flew around the outcrop area to survey its boundaries, which was easily done because the black rock was well-exposed. The peridotite was a crescent-shaped outcrop, about one and a half kilometers long and three-quarters of a kilometer wide. The lake was in the center. We landed on a mound of black rubble, near an exposed contact with granite, and spent thirty minutes prospecting along its borders, looking for rust spots that would indicate a concentration of nickel-bearing sulfides. We found nothing of interest.

Then, we examined the lake. The reason for its strikingly different color was easily seen. Theoretically, peridotite is a rock composed of only two minerals, but, in fact, it often contains considerable amounts of accessory minerals. This peridotite contained about 15 percent dark mica. Freeze-thaw weathering had released the mica from the rock, and it had collected on the lake bottom as silt. High winds caused the micaceous silt to go into suspension. The flat reflective surface of the mica made the water appear opaque black, even though the water of the lake contained a smaller percentage of suspended particles than the pastel-colored lakes.

These are the rainbow lakes.

CHAPTER 14

The Margin

The Hudson Bay trading post is near the mouth of the Perry River, where it empties into the Arctic Ocean. It is at the edge of the world. There is a bit of civilization there, but it is of dubious quality. We had not intended to visit, but at eleven o'clock in the morning it was only ten kilometers beyond the end of our traverse, and we decided to drop in to have a cup of coffee with our lunch.

We created a sensation when we circled the post looking for the wind direction and searching for a landing place along the shoreline. Heads popped out of tents, and groups of men started trotting toward the post. When we got lower, we saw that many of what we thought were tents were tarp covered sleds, piled with Inuit belongings that were not needed during the summer. They were stored on the highest part of the boulder beach. The post itself looked newly built, and was located about ten meters above high tide in the middle of a field of boulders. Paths had been cleared from the factor's house to the warehouse and dock, but otherwise one traveled by jumping from boulder to boulder. We landed at the water's edge where sand had filled the space between the boulders, and where the resiliency of the chopper's rubber pontoons compensated for the uneven surface.

While the rotors were slowing, a crowd gathered to watch, spreading out on both sides of the factor, who had come down to meet us. We stayed in the chopper until the rotors stopped turning. Then, I got out of the bubble, ducked under the blades, and went up a small slope to greet the factor. He was a young man about twenty-six years old, of medium height and slight build, bareheaded, and very white when compared with the darker crowd that surrounded him. I held out my hand: "My name's Ron Seavoy. I'm with International Nickel Company."

"E'm Leal Embertake en em gled to new'ya," he said in unintelligible English, though not in a harsh accent.

"What's your name?" I said, "I'm sorry, I missed it."

"Leal Embertake."

"Glad to know you. We thought we'd drop in and say hello."

"Gud, On't ya cum up fer coffee," he said.

"Fine, we'd never turn that down, but hold it a second." I turned and yelled back to Fred, "We're going up to the house to put on coffee. Come up as soon as you can."

"Are ya from Beechey Lake?" he said.

"Yeh, how'd you know?"

"I heard on one of my radio messages that you were in the area."

As soon as I got inside the building, I commented: "New buildings aren't they? How long they been up?"

"About three years. Moved from the Perry River island because it had no fresh water." I was beginning to understand him. He did not speak in a slurred slang, as I had first supposed, but in a very British accent. Yet, it was one I could not place. I asked, "How long you been here?"

"Only a year."

"Kind of a lonely spot. Desolate country in fact! Where were you before this?"

"Chimo and Payne Bay."

"How long ya been with the Bay?"

"My sixth year."

"All in the north?"

"Yes, indeed. All in the Arctic."

"How long ya going to stay?"

"My last year. I'm going to Australia next."

"That's a little more sensible. They got great spaces of nothing out there, too. I'll probably get there someday, if I stay in this business." Then I saw his shotgun against the radio cabinet, and said, "Goose shooting ought to be good around here. Do ya get many?"

"No, I've not had much of a chance."

"How come? On the way here we saw lots of flocks V-ing up and heading south."

"It's about time they do that. There's lots in the area. The Natives get them in season, but there's not too many near here, and I never seem to get far enough away to get any."

"Do they get them with clubs when they're molting and can't fly?"

"No, mostly they trap or snare them."

"They don't shoot them?"

"Indeed, no!" he said emphatically. "Ammunition costs too

much money for Natives. They don't get too many."

"Huh? I'da thought different by their numbers." At this point, Fred came in and I said to the factor, "This is our chopper pilot Fred Leegirt." Fred extended his hand and the factor repeated his name. This time I caught it.

"I'm Neal Timberlake." Fred nodded and they shook hands.

Then, I said to Fred, "We're in luck, we get a cup of coffee."

Fred said, "We were hoping for one. I brought our lunches." Neal slid a primus stove out of the corner, and lit it. While it was warming, he got out the coffee pot and filled it. While it was coming to a boil, we made conversation.

"Neal," I said, "where'ya from? I can't quite place the accent."

"South Africa, in Rhodesia."

"Bulawayo?"

"Yes," he said with a look of complete surprise on his face.

"I'll be damned, I'm from Chicago and Fred's from New Zealand. Well, what the hell do you know about that! I'm with a couple of colonials, only don't use that word in Canada, or have you found out?"

"Very early in my stay here. It's asking for a fight."

"Asking? You'll get one!"

Fred asked, "Did you ever get stuck with that handle in England?"

"Oh, indeed, yes, but we really are colonials you know, so they're quite right."

"Not so for New Zealanders," Fred said. "We fought in the wars in our own armies."

I turned to Fred and said, "He plans to go to Australia next year."

"Where are you leaving from?" Fred asked.

"I don't really know. I haven't yet decided; maybe Vancouver or perhaps San Francisco. I want to see a bit of the States before I leave."

"You should see 'Frisco," I said.

"Yeah," said Fred. "It's a real town: a true city. It's got a real downtown, not like Los Angeles where I've been only once. That city's a jungle of highways going nowhere."

"I want to get there before I leave, but Los Angeles has such poor connections out of it."

I backed up Fred's opinion: "Unless you get a car, don't bother going there. Spend your time in Vancouver or 'Frisco."

"Why ya going to Australia?" asked Fred.

"I'm on my way around the world, I haven't been there yet, and it's on the way. My travels have taken a bit of time, but I'll get there."

"How's the tide here?" Fred asked, "I've got the machine sitting on the shore."

"It's only a meter, and it's high tide now. I don't think you have anything to worry about."

"You said you were at Chimo," I said. "They have some tides up there, don't they?"

"Extraordinary, indeed! They are something to see. When I was at Payne Bay, we had fourteen meters. That whole coast has very high tides, indeed."

"I flew over Payne Bay in '57," I continued, "in the early spring, just as the ice was going out. The bay was jammed with ice floes, and you usually can't see them move. They just slowly shift positions. Well, I could see them being eddied about by the tide, as if they were bits of driftwood in a torrent."

"Yes, they made a fearful noise when they smashed into each other. That happened all the time when the tide was coming in, and if you were on the flats when it came in, you could hardly escape by running full speed to the shore."

"You at Payne Bay and Chimo both?" I asked.

"Yes, indeed. I spent a year at Chimo, and six months at Payne. I was learning the trade then."

"Were you at Chimo in '57?"

"I think so, but . . ."

"When the Ungava rush was on?"

"I believe I was. Yes, I think so. There was a fast pace there for a while in the early spring, before the ice went out and everyone moved north."

"Did you spend much time in Chimo town or were you pretty much at the post downstream?"

"I got up to town a few times, when I had to. I didn't like it up there. Mostly, I was at the post and away from the activity, but I did get up there several times when I had to see D.O.T."

"Do you remember 'Chez Paulo'?"

"What?"

"Chez Paulo, that so-called hotel at the airstrip, where those going north stayed because they had no choice. It was Baffin Air's converted radio shack. It was run by Paulo Kelly, a wild little Frenchman, who used to get on the phone and yell at the top of his voice in French, then suddenly switch to English and back again to French when he wanted to be emphatic—which was most of the time."

"Yes, I know where you mean, but mostly I remember the one meal I had there, because I had to spend the night. They were good to me. They didn't charge me, and I wouldn't have paid. Horrible! I got greasy beans for breakfast, and I'll remember that for awhile yet."

"Now that's God's truth," I said. "I think the cook was half French and half Eskimo [Inuit] who inherited the worst of both. He cooked as if he'd never been civilized, and he probably wasn't. The prices matched. Breakfast was $2.50, and lunch and supper were $5. The sleeping conditions were just as bad." I continued, "Paulo charged $5 a night for use of a double bunk, and you had to provide your own sleeping bag. It was a bit crowded, and there was nothing to do but wait until your plane arrived. I was going to Wakeham Bay. Ever been there?"

"No, I was only up to Payne."

"Did they have the airstrip when you were there?"

"Airstrip? I didn't know they had one."

"I guess it was built after you left. A year later they did lots of drilling on an iron ore deposit. Can't think of the name of the company that did it. 'Arctic Ore Company'? Does that ring a bell?"

"Was it 'Oceanic'?"

"That's it! There were plans to put in a mine and mill and then ship the concentrate to an open water port in Newfoundland where it could be transhipped to Europe for twelve months of the year, instead of only four from Payne. There was some German company interested in it—Krupp, I believe. There's a spot near the mouth of the bay, at a cliff, where a little work could've made a dock where ships with twenty meter drafts could be moored at low tide."

"I just don't know the spot. We were located at the head of the bay, and Oceanic was just getting started when I moved out."

"I imagine the weather's as bad here as at Payne. When we were camped at Tree River, we got some nasty weather: two days of winds and a cold, miserable, dripping fog. I'll bet you get many days of sixty kilometer wind with matching cold during the winter."

"No, surprisingly! During deep winter we don't have many days with winds over ten kilometers per hour, in fact, very few."

"What about the fall and spring when the seasons change?"

"Oh, we get weather then, all right."

"What about winter temperatures?"

"Oh indeed, we get some low ones. Last year we had five weeks when it was constantly forty-five below."

"Bother you?"

"No, not much. The winds at the change of season, like what's coming up soon, are sometimes hard to take. Ya know, five or six days with gale force winds with rain, sleet, fog, and snow. Altogether miserable stuff."

While I was refilling my cup, I noticed a semi-circle of Inuit behind where we were sitting. They were smiling, listening, and staring. The door had been left open, and they had come in. One was conspicuous because he wore a pair of steel rimmed spectacles, and another because he wore a pilot's hat. All of them wore various combinations of Western clothing, plus the inevitable rubber boots. As we continued chatting, some squatted but they always beamed a broad grin when we turned their way. They all looked healthy, and in good spirits, so I asked Neal, "Are there lots of seals around here?"

"Normally, they're not too plentiful, but this year there seems to be a good many, but for some reason the Natives don't go after them much."

"What do they live on?"

"Fish. That's about the only thing that will provide a steady food supply. They shoot the odd caribou in the summer. A caribou is worth a bullet, but they don't shoot any great quantity like they do at Contwoyto. They also get a whale or two in the spring and fall. But, it's mostly fish."

"What kind of fish do they get?"

"Char and trout."

"What about salt water fish?"

"There aren't many big ones."

"You mean their diet's mostly fresh water fish?"

"Oh yes. They do their winter fishing on the larger inland lakes. There's no abundance of fish, but they catch enough to get by."

"I'll be damned! What about cod? There's cod up here isn't there?"

"Tommy cod, lots of that and lots of herring, but they're all small and bony, and the Natives have to be pretty close to starving before they'll eat them."

"How frequently are they near starvation?"

"They get rough times, but they don't usually starve anymore. When it gets too bad, the Mounties fly in a load of rations, if they know about it. But, sometimes it's a tight race, and sometimes it's too late. Three years ago, a band of sixteen starved in the Pelly Lake area."

"How do they catch fish? Nets or spearing?"

"Nets, which they spread by boat in the summer, and under the ice in the winter."

"How's the char situation here? Fished out near the post?"

"Yes, indeed, and besides the water's a bit muddy and there aren't many, but in small rivers the fishing's fair enough at this time of the year. They're beginning to run now. They go up river during the fall and stay in the lakes during the winter. That's when the Natives go inland to fish."

"Fred and I stopped yesterday where a small stream enters the sea, and we caught lake trout in salt water, or what was probably brackish water. Do they catch any trout like that here?"

"They catch a few trout in the nets. The ocean along this coast isn't very salty because of the fresh water coming from the Perry and Ellice rivers; and the low tides preventing mixing."

"Do the Eskimos store food for hard times in the winter?"

"None a'tall."

"They could smoke fish couldn't they, or dry them?"

"Yes indeed they could, but it would take great effort. This is not like the Bathurst Inlet post, where they have enough vegetation to build a good smoke fire. This is very barren country."

"What do they use for fuel during the winter?"

"Mostly nothing. Sometimes they buy a little fuel oil, and they gather what gas they can when an airplane comes in to refuel, and they use what's left in the bottom of empty barrels, but this is a very poor group of Eskimos."

"Are we the first chopper in here?"

"I think so, but I've only been here a year. You're the first since I've been here."

"What about boats? I was told that Eskimos around here don't make kayaks anymore. Is that true?"

"Yes indeed. The only place kayaks are still used is in the east, and there every Eskimo has one, and they use them all the time. They haven't made any here for a long time. They wouldn't know how to use one."

"Do they use those wooden boats made by the Eskimos at Tuk?"

"Not if they can help it. They're too heavy and clumsy, and not well-made. They prefer square-sterned canoes with kickers, and they'll save to trade for them. I doubt if I'll sell one of those Tuk boats. I've had them on the rocks for almost a year, and haven't sold one. You saw them up there when you came in. They're really not very good boats."

"They look awkward, compared to square sterns. Is this post purely for the fur trade?"

"Yes, that's all."

"Only white fox?"

"Mostly fox, but some silver seal and wolf. We also get an occasional wolverine, and a bale or two of summer caribou skins. We send the caribou skins north to Cambridge Bay, and sell them to the Natives for winter clothing, along with caribou sinew for thread."

"What! Eskimos buying skins to make clothing! Amazing! Why only summer skins?"

"Winter skins are too heavy; hair's too long; makes them too warm, and they sweat and then freeze. Winter hides are for covering tents."

"What do you do with the wolf and wolverine skins?"

"We sell a few to the Eskimos, but most of them go south where the curious buy them."

"What do the Eskimos use 'em for?"

"Wolverine's the best fur there is for lining around hoods. The frost from breathing won't freeze to it. When it accumulates, you just brush it off. They also use it around the ends of the sleeves, and around the bottom of parkas, wherever moisture is likely to collect and freeze from sweating or breathing. Winter wolf fur, that's heavy and white, is second best for this."

"Are the silver seal and fox fashion furs?"

"Yes, we ship all of them out."

"Is the silver seal related to the white cub seals that are killed off Newfoundland and Labrador?"

"Yes, indeed. That's the cub of the silver seal, but we don't get them here. The Natives take only two year olds. Actually, the fur at this stage has more value. I don't get many."

"How many Eskimos trade here?"

"Right now ninety-four, but they're scattered along the coast for 300 kilometers, and they're never more than a few here at a time."

"Can ninety-four people provide enough business to make this post pay?"

"Not now, but this post was built for the long run, gambling that the white fox will come back into fashion. It was also built to clear up an embarrassing situation. You remember the case of the trader up here who got drunk and stabbed an Eskimo? They had a big trial over it. At least, a lot was written about it."

"Vaguely. . .it rings a faint bell. Did he kill him?"

"Took three days for him to die, but die he did. Well, this was the post where it happened, but it was located on the island then. Well, this trader was an independent, but supplied by us. We were, you might say, persuaded by the Department of Northern Affairs to buy him out, and he now lives in that building up in the cleft in the rocks." He made a vague gesture to the area behind his back.

"You mean they didn't put him away?"

"What could they do? That would have been a vacation. You know, he was a half-breed, little better than the Natives he dealt with. Now, he's retired and out of the way, and we took over to clean up a botch. Let me give you an idea of what he was like. Do you know how he ordered his supplies? Really quite quaint: six cases of 'Fragile,' four cases 'Do Not Freeze,' eight cases 'Handle With Care,' and one case 'Product of England.'"

"I remember the case," Fred said. "It happened about three years ago, didn't it?"

"Actually four, but the Mounties didn't get around to doing much about it for a year."

"So, that's the guy," Fred continued. "They couldn't have put him in a worse spot than here! I was looking at the number of gas barrels you have here," Fred continued. "You don't have many. You don't get many planes in here, do you?"

"Only a few. We get a few drums for our plane, which stops in once or twice a year to take out furs and bring the boss around inspecting, and some of the airlines have the Hudson Bay ship leave a few for emergencies. But, that's about all."

"You'll have a lot dropped off this year, ya know."

"I heard about it."

"There's a helicopter mapping project by the Geological Survey scheduled for next summer."

"I'm told they're going to leave ninety tons of gas drums, but it could be more."

"When's the boat expected?" I asked.

"I'm told about the first or second, but I'm not sure which ship is coming. This post usually gets supplied by the Hearne, out of Tuk, but some years it's a ship out of Churchill. Both have gas aboard. That which isn't dropped here will go to Cambridge Bay."

"When's the plane coming for the kids who want to go to school?" I continued. "How many you got here?"

"I don't exactly know when it's coming, probably sometime this week. We'll probably have the same number we had last year—fourteen."

"Fourteen's a good number, compared with those that go from Contwoyto Lake."

"I guess it's because they're better off there than here, although, that seems unlikely. It's up to the children, you know, whether they want to go. Eskimo parents don't force them to do anything."

"Where're the kids that are going? I don't see many here."

"Up and down the coast. The plane'll stop and pick them up wherever they're camped."

"Do they learn anything?"

"Not much. Oh, I suppose it helps some. They learn a bit of English, but mostly they consider it a vacation."

"How long they been going to school?"

"Last year was the first time, so maybe there won't be so many this year now that it's been tried. I really don't know."

"You speak Eskimo don't you? Don't they tell you who's going?"

"I don't really speak it. It's a very difficult language, but I get along in it so I know what's going on, so I can tell the Mounties or D.O.T., and can trade furs. But, they don't tell me about these things, even when I ask a direct question. They'll just sort of let me know that maybe Joe Kenoyatuk's son might be interested in going, and tell me where he's camped. The plane will stop and maybe his kid will get on, or maybe they just want to see the plane land."

My third cup of coffee was finished, and Fred and Neal had had their second. The pot was dry. We got up to leave, and Fred said, "I think I'll go down and check the oil. I'll probably have to put in a quart."

"Okay," I said, and then turned to Neal and said, "What's that other house over there?" pointing to a white canvas-covered cubicle.

"That's the priest's house, when he visits. He comes from Cambridge Bay."

"RC [Roman Catholic]?"

"Yes, but most of the Natives here go to the Anglican priest, at Coppermine, when they want to get formally married. They always combine marriage with a trading journey. Eskimos can't understand how a man can be a specialist in religion, so they don't have much use for the RC. Their life depends on doing a lot of things reasonably well. A priest who specializes, and is celibate, is unnatural. The Anglican priest is married, hunts, fishes, builds things, and gets things from the outside, and they understand this."

I started toward the chopper by the long route, via the dock. On it was a pile of boulders, and I said, "That's soapstone, isn't it?"

"Yes, indeed."

"Where's it going?"

"Cambridge Bay. They asked me to get a ton, but I could only get 500 kilos. It's going there when the boat arrives. Next year, the stone will come in by boat from Southern Quebec."

"Don't they carve here?"

"Not a'tall. They've never done it 'til now, and they're just starting at Cambridge. They brought in an eastern Native to show them how."

"Are you going to start them here?"

"Not this year. I think my replacement will be from the east, and he'll bring some models so the Natives can learn."

"You know, Eskimo carvings are hot items in the stores."

"So I'm told. They want to start it here as soon as they can. They've made money on it in the east, where it's done naturally. Here, we've got to teach them just like we've got to teach them to run an outboard. All the Company cares about is that the Natives do it."

When I heard Fred start the chopper, we began hopping from boulder to boulder toward the shoreline. Just beyond the arc of the rotors, I extended my hand, and said, "Thanks for the coffee. It's been a pleasant chat. I doubt if we'll see you again, so good luck."

"So long, and do drop in again."

CHAPTER 15

The Wolf Hunt

A week after we visited the Perry River post, we broke camp. The weather had rapidly cooled during the week, and we had three mornings of dripping fog, one day of off-and-on drizzle, and one day of rain. We could fly in the intermittent misty weather, provided there wasn't a high wind, but we could not fly in rain or in dripping fog, nor on windy days. Wet and windy conditions would become increasingly common during the unstable weather that precedes the change of season. Flyable weather might last another week, or the weather might steadily deteriorate into a prolonged period of high winds, freezing rain, and snow. It was time to move south. Keith made the decision on August 26th and the next day he brought in both fly-camps to provide labor. But, it wasn't until August 28th that we broke camp.

On the 26th, we put a partial load aboard the Beaver, and Keith joined Jim Shaver to locate a new campsite. He found a good site, 300 kilometers almost due south, at the western end of a shallow sand-bottom lake that was part of the Thelon River. After the plane was unloaded, they went to Yellowknife (400 kilometers due west), where Keith chartered another plane to help move camp and arrange for several loads of chopper gas to be flown to the new campsite. He also made reservations for the entire crew on outbound flights to Edmonton after the season ended.

The next day (August 27th), Shaver left Yellowknife with a load of aviation gas for the new base camp, then flew north to join us. He arrived late in the evening, and we loaded his Beaver in preparation for breaking camp the next day. In the morning, Shaver left with two men and a tent so that, when the chartered plane arrived at the new campsite, labor would be available to unload the gas drums. After delivering them to the campsite, he returned for another load and overnighted with us. The following morning, on the 6:30 a.m. radio communication, we were informed that we would not get a chartered plane until the next day. We had enough gas and the weather was flyable, so we made one more chopper traverse west to Contwoyto Lake, but found nothing.

The scheduled morning radio, on the 28th, confirmed that the chartered plane would arrive within three hours. This meant we were free to begin our exploration traverse as soon as we performed our share of the labor of breaking camp. We helped load Shaver's Beaver, and delayed starting our traverse until the chartered plane arrived, so we could help load it. It came at ten o'clock in the morning, and both of us left an hour later.

On the trip to locate the new campsite, Keith had discovered a gossan twenty-five kilometers south of camp. We sampled it and continued southeast, inland from the west shore of Beechey Lake. After leaving the Benchmark Lake camp, we began seeing pairs and bunches of caribou every kilometer or two, so that, in an hour's flying time, we counted eighty-three. Dispersed animals were beginning to form herds and move south. Then, coming over a hill that formed the lip of a valley draining into a large lake, we saw a pack of six wolves following a small herd. In three years in the Arctic, this was only the second time I had seen wolves, both this year and both in packs. They were not yet in winter white, although the white undercoat was beginning to show through the coarse, black-tipped summer hair.

Fred immediately set the chopper down, and jumped out to get his 7-mm rifle. We never carried the rifle, except on camp moves. As the pilot, he was privileged to carry most of his personal possessions with him, and his rifle was in the same class as his flight log, personal papers, and fishing rod. It was, however, lashed on the outside cargo rack. While Fred got his rifle from its case, I dug into his briefcase and found the box of shells. All the wolves had trotted off while he was putting five shells in the magazine.

By the time the gun was loaded, there were no wolves in sight. We decided to track and kill one, if we could drive it in the direction we were going. I would ride shotgun, holding the rifle, tracking the animal, and navigating, while Fred flew fifteen to twenty meters above the ground. We started chasing the pack. They immediately scattered, so we singled out a large dog and began the chase. He persisted in going east, so we circled until we located another animal going south.

After five kilometers, he began looking back at the great bird that would not stop following. We usually managed to stay within 300 meters of him, except when he got among boulder patches where we lost him. This happened a couple of times. Then, we went up in the air, and hung there until we saw him move. At the end of eight kilometers, his tongue was lolling in his mouth, and we backed him up against the shore of a lake where he paused, nearly exhausted, and lapped up some water. Fred set the chopper down as quickly as possible, about 120 meters away. I screwed down the stick, and tightened the collective, while Fred grabbed the rifle and jumped out of his door. He ran to the rear of the chopper, flopped into a sitting position underneath the moving rotors, and squeezed off three shots before the wolf got out of sight among large boulders.

When he handed the weapon back to me, he noticed that the 200 meter sight was up and this explained why he missed. By the time we were seated in the chopper, we had taken our eyes off of the wolf, and we had to circle for three or four minutes before we saw him again, loping up a hill a kilometer away. When we resumed the chase, he broke into a run. After another kilometer, he slowed and then stopped, with his back to the base of a small cliff. But, before Fred could get out of the chopper, he was off again. However, he was vastly slowed, and we practically hovered over him to keep him moving, until he used up his last reserve of strength. It took another kilometer. Then, he turned to face us along the shore of small lake, where he feebly lapped up water, all the while keeping his eyes directly on us.

Fred landed and, as soon as the chopper was settled, he unbuckled his safety belt, while I again screwed down the collective and tightened the stick. When he was outside on the ground, I handed him the rifle. He again ran behind the chopper, squatted under the boom and took aim. I watched the wolf through the bubble, still strapped in my seat. Slowly, the wolf began moving along the shore at an exhausted trot, and he was beginning to gather speed when Fred fired. The wolf tried to leap ahead, but his hind legs would not obey and he spun into the water with a splash. He struggled to get up, then collapsed and lay still.

I reached over, and pulled the carburetor feed switch. The engine sputtered, and died. I loosened my safety belt and got out of the plane, while Fred came around and turned off the generator and master switch. I took my geology hammer from the equipment rack, and we started toward the carcass. As soon as we got clear of the tail prop, I turned to Fred and said, extending my hand at the same time, "Congratulations on one shot."

He shook hands with a big smile on his face, and said, "Thanks, if that sight hadn't a been up, I'd got him the first time."

We walked in silence for the remaining fifty meters, and then I asked Fred if he had his skinning knife. He stopped short, and felt in his pocket and along his belt, and then remembered that he had packed both knives in his suitcase, and they were being moved by plane. He said, "Oh, I guess we won't bother with the skin. It's probably not any good this time of year."

He was right, but I said, "No," implying that what was the use of killing a trophy, if the skin was not taken? "I have mine and I'll skin him."

The carcass was in the water among shoreline boulders. Only the tip of an ear and the long hairs of his tail were visible. I stepped on an underwater rock and grabbed the tail, while Fred was saying, "Let me, don't get your feet wet." But, I'd already grabbed a hold of him. I pulled him on the muskeg, and we looked at his outstretched body. It was a two year old dog, in prime health, just approaching physical maturity. From the tip of his nose to the end of the tail, he was about one and a half meters long, and his paws were as massive as the palms of a raw-boned Newfoundlander. After we had seen him, Fred grabbed his tail and dragged him to high ground where he could be skinned. On the way, I asked Fred how much he weighed. He ventured about twenty kilos or a bit more, to which I agreed.

The bullet had entered the lower rib cage, and traveled half the length of his body cavity, before emerging through the chest on the opposite side of the body, just below the heart. Where the bullet came out, it nicked the right front leg, barely cutting the skin. The skin was thin, and the fur light. However, the undercoat was beginning to turn white. I did the skinning, and it was an amateur job. I left a lot of meat and fat on it, although I avoided cutting it.

When the carcass was stripped, I decided to take the head because it had a perfect set of teeth. I cut out the tongue, sliced off the nose and ears, and then partially cut through the massive neck muscles, but the knife had difficulty severing the tendons at the base of the skull. I finally managed this by placing a large flat rock under the neck bones, and taking the geology hammer to the base of the skull. It took more blows than I had anticipated to shatter the neck bones and mash the tendons, because I had to aim at an obscure mass of red flesh where I thought the bones were. The next day, I boiled the skull in a pail of water, stripped off the flesh, and boiled it again, after adding some laundry detergent containing bleach. Now, the whitened skull, with its glaring, devouring teeth, stares in cold ferocity from the top of my bookcase.

About thirty kilometers south of the wolf kill, we broke out of the overcast into bright clear weather of twenty-two degrees. We had passed through a stationary weather front, which meant we would be erecting camp in warm sunshine. This is always a blessing. We arrived at the new campsite after three short stops to sample small gossans.

According to the map, the new camp was eighty kilometers north of the treeline but, since we were down in the Thelon valley, there was considerable protection from northern winds. On the southern exposures in the lea of bluffs, groves of stunted trees grew on well-drained patches of sand. These patches were never more than a few acres in extent, and the trees growing on them were a maximum of two meters high, but they were a welcome sight.

As soon as we arrived, we helped set up the tents, put all of the equipment under cover, and most of it in place. It was the end of an arduous day, as all camp moves are. We crawled into our sleeping bags at 10:30 p.m., and slept the sleep of the dead.

CHAPTER 16

The Search for Precambrian Life

Keith remained in Yellowknife during the move to the Thelon River base camp. On the morning radio schedule, he told us he would return the next day, but we would have to meet him halfway. He would come by chartered plane to the weather station at Reliance, at the eastern end of Great Slave Lake. He would bring aviation gas to refuel the chopper and leave two barrels there in case of an emergency. Our new camp was 200 kilometers east of Reliance, and we would make a prospecting traverse there to pick him up.

The east arm of Great Slave Lake lies in a deep basin. The lake is only 156 meters above sea level, but the rocks that form its southern rim average 425 meters in elevation. The rim rocks are hard granites that are 3 billion years old. The rocks in the basin are 2 billion years old, and are much softer because they have not been intensely metamorphosed. Within the basin are several ridges of glacially scoured rock that form gentle arcs tens of kilometers long. In several places along their length, these ridges have scarps which rise over 240 meters from the water's edge. The highest crests of these ridges are blunt peaks that are nearly as high as the enclosing rim. The rocks that form these persistent ridges are of two kinds: the thick, Pethei limestone; and diabase dikes and sills, like the dikes and sills that formed the linear ridges, and sugarloaf knobs in Bathurst Inlet.

If we had had time, the trip to Reliance would be an opportunity to visit outcrops of Pethei limestone. Some beds within this formation contain algal structures called stromatolites. Algae are microscopic photosynthetic plants that live in water. Most are free floating, but some species live in colonies. Stromatolites are the colonial form. They are frequently preserved in rocks, but deep burial, high temperatures and pressure, have destroyed the cell walls of the algae that built them, although their colonial structures survive intact. Stromatolites still exist. In North America, they are found in the lagoons of the Bahama Islands off of the east coast of Florida, but the most spectacular surviving stromatolites are in Sharks Bay, in Western Australia, 650 kilometers north of Perth. In both places, the stromatolites are identical to the 2 billion year old fossilized stromatolites we observed in the Pethei limestone west of Reliance.

Outcrops of the Pethei limestone are among the best sites in the world for collecting evidence of Precambrian life. The Cambrian Period began 560 million years ago. Its beginning was marked by the appearance of animals that had shells that were easily preserved, and could be seen without magnification. The 3.5 billion years before the evolution of shelled animals is called the Precambrian Period. It constitutes most of the geologic record. By way of comparison, the oldest dated rocks in the solar system, as measured in meteorites and moon rocks, are 4.5 billion years, and the oldest dated rocks on earth are 3.8 billion years.

The lack of fossil shells does not mean that Precambrian life was scarce. In fact, it was vastly abundant, but most of it was microscopic. Recently, several teams of geologists in Canada, the United States, South Africa, and Australia have found rocks that contain abundant microfossils of single and multi-cellular plants and animals, that can only be seen under high magnification. Canada has several places where Precambrian microfossils are abundant. The best known is Schreiber, Ontario, the name of a railroad siding east of Thunder Bay, along the north shore of Lake Superior. These fossils are carbonized cell walls of algae that are preserved in chert. Chert is a variety of flint. It is a siliceous rock that resembles glass. If Precambrian chert is not deeply buried, and has escaped the high temperatures and pressures of deep burial that have metamorphosed most ancient rocks, it will preserve the remains of the most delicate and fragile one-celled plants that proliferated in the quiet waters of marine bays and lagoons.

Previous to the discoveries of microfossils, there were many indications that primitive life existed, although there was no positive proof, because there were no identifiable remains of algae cells, siliceous spines, or limestone shells that are the usual fossils of Postcambrian plants and animals. Nor were there any vertebrate animals, because the first vertebrate animals did not appear until 450 million years ago.

One of the most common indirect indications of primitive life is the abundance of graphite beds in the earliest Precambrian rocks. Unfortunately, graphite beds have few internal structures that indicate an organic origin, yet how else could carbon have accumulated in such large quantities? For carbon to be preserved in such large quantities, microorganisms had to be immensely abundant in tropical seas, and there had to be many favorable environments. After algal remains accumulated, heat and pressure from deep burial converted the carbon to graphite. Many geologists believe that graphite beds are ancient coal beds composed of the remains of single cell algae. Some coal beds, composed of vascular plants (250 million years old), have been converted to graphite by the same processes. Another argument for the abundance of Precambrian life is the presence of chert beds in many Precambrian sediments, assuming that some of the silica in sedimentary rocks was concentrated by microorganisms, as it is today.

However, the most positive indication of early life is the presence of stromatolites in Precambrian limestones, wherever they are found on earth. Stromatolites have been given many names because they occur in rocks of vastly different ages, from over 3.4 billion years ago to the present. When Precambrian stromatolites were first described, they were called Collenia, while those found in Postcambrian rocks were usually called Cryptozoon. Stromatolites are extremely good evidence for the existence of Precambrian life, especially when living forms have identical shapes to those that are 3.4 billion years old.

Colonial algae that lived a free-floating existence in the open ocean have spherical shapes. They are called oncolites. They were the most abundant Precambrian form of colonial algae, but they do not preserve well in the geologic record because storms disintegrated them. They did not incorporate lime, ($CaCO3$), into their spheres, because it would have caused them to sink. In order to survive, they had to float in the sunlight zone at the ocean's surface. They did not need protection because there were no predators at the time. Oncolite fossils are occasionally preserved, if they died in quiet lagoons and were covered by a continuous rain of fine-grained lime or silica. One such lagoon is preserved at Taltheli Narrows, where the east arm of the Great Slave Lake joins the main body of the lake.

Shapes of algal colonies depend on the energy level of the water they live in. Colonies living on tidal flats form mats, because a mat is the best design to resist desiccation. Mats are made of gelatin secreted by algae, that have string and rod shapes. They form a felt that imparts strength to mats, and the gelatin conserves enough water for algae to survive the twice daily exposure to air, during low tides. Some colonial species enhanced their ability to survive on tidal flats by synthesizing lime, in order to add weight to the mat. The weight helped prevent algal mats from being ripped up, and shredded during storms.

Stromatolites evolved other shapes to live in high energy environments. The high energy environments are generally in tidal zones with strong currents. Their survival requires them to be strongly anchored to the bottom, in order to resist storm surges. Algae living in stromatolite colonies in high energy zones, whether in Precambrian or contemporary, synthesize lime in order to build heavy biogenic anchor stones that are embedded in the bottom mud. The anchor stones resist dislodgment by currents, because of their heavy weight. The large size of some of these stromatolites makes them highly visible when they are exposed by erosion.

If anchor stones are not undermined by tidal scour or quickly buried by sediments, stromatolite colonies can live for many years. Whatever their shape, however, Precambrian colonial algae had to perform four functions to survive: 1) gather an optimum amount of light; 2) have an optimum surface area in contact with water to extract mineral nutrients; 3) resist being torn apart by moving water; and 4) grow upward to stay above the level of accumulating sediments. Contemporary stromatolites perform these same functions.

The most common form of stromatolites preserved in the Pethei limestone are inverted cones forty centimeters high, and about fifteen centimeters in diameter at the top. All stromatolites grow upward, and many grow outward. If a colony failed to synthesize lime fast enough to escape being buried by sediments, it died of suffocation and became a fossil. In environments of sediment accumulation, only ten to fifty centimeters of the crowns of stromatolites protrude above the bottom mud. This is the living portion of the colony. Sometimes the crowns are hemispherical, like pencil erasers, overturned thimbles, or overturned fruit bowls, and, at other locations, where they survived long periods of rapid sedimentation, they look like tree stumps.

Colonies that synthesized lime fast enough to exceed the rate of sedimentation have deep biostone bases that gave them great stability in turbulent water. In the visible cliff faces, at lake level, in the east arm of Great Slave Lake, giant stromatolites are exposed in

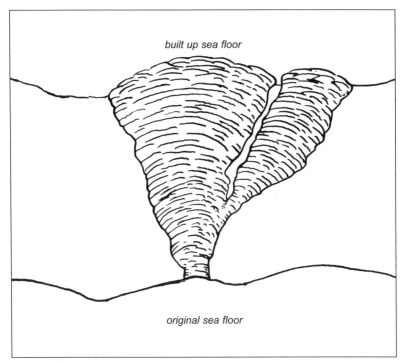

built up sea floor

original sea floor

Stromatolite.

cross-section. These fossil stromatolites are up to three meters high, and look like trunks of palm trees snapped off by hurricane winds. In the process of building upward, the gelatinous mats that sheathed the crowns of stromatolites trapped fine grains of silica sand on their surfaces. This produces an onion ring appearance in fossilized stromatolites. The onion ring structure gives stromatolites of all sizes their distinctive appearance.

There is now sufficient evidence to say that a high percentage of Precambrian limestones, as well as a high percentage of recent limestones, are composed of biogenic minerals. In the oldest rocks, algae produced these minerals, but in recent rocks, corals and mollusks usually produce them. A high percentage of recent limestones are really mass fossils that have had the direct evidence of their biogenic origin destroyed by the fragmentation of shells followed by recrystalization. Then, as now, the preservation of complete fossils, whether stromatolites, corals, or mollusks, is very much the exception rather than the rule. The association of stromatolites with Precambrian limestones is good evidence that there was an abundance of life in Precambrian seas. This conclusion changes our concept of Precambrian times. The usual concept is that life was scarce. It was scarce on land, but in tropical seas it was immensely plentiful. Most of it consisted of a very limited number of species of microscopic plants and animals and jellyfish that lacked shells.

We began our traverse to Reliance at eight o'clock in the morning, which was a bit early because every day the sun was sinking lower on the horizon and, on cloudy days, good visibility did not begin until about nine o'clock. But the weather was clear, with a warm south wind. We hadn't worked in summer weather like this for two weeks. Besides, we wanted to arrive in Reliance with time to spare. We did not know the exact time Keith would arrive, because he himself was not sure when he would leave Yellowknife. If he got an early start, we wanted to be there waiting for him.

Three-quarters of an hour after leaving camp, the sky began to get hazy and, at the next gossan we sampled, we understood why. We smelled wood smoke, although we were still in the barren lands. The treeline was seventy-five kilometers away. The ground temperature was uncomfortably hot, because the smoke created a greenhouse effect, raising the ground temperature to nearly thirty degrees. It was even hotter inside the plastic bubble of the chopper. The heat, plus three men in the cockpit, would make the return trip even less bearable.

As we approached Great Slave Lake, we were flying about 100 meters above the ground. This was about twenty meters below our normal prospecting altitude, but we had to come down because smoke was obscuring many surface details that aid navigation. Suddenly, five kilometers from Reliance, we were suspended 375 meters above the earth looking down on the flat, blue surface of Great Slave Lake. We had crossed the McDonald fault scarp that forms the southern rim of the east arm of the lake.

The McDonald scarp rises 275 meters from the lake in half a kilometer. At this time of the year, with the sun getting lower in the south, the scarp was in deep shadows and the sensation of being suspended in the sky was heightened when we dropped into its shadow, while descending to Reliance. Added to the feelings of suspension was a feeling of loneliness because Fred throttled back the engine to near silence. The deafening throb of the engine ceased, and it seemed that nothing could keep us airborne. But, our air speed remained the same, and we lost altitude gradually because we were riding the top of a great unseen cataract of air that flowed over the McDonald scarp and into the basin. We floated on top of it, and it gave us enough lift, as we descended to lake level, so that we glided almost the full distance to Reliance.

We arrived just after noon, but Keith was not there. The weather station operator informed us that he would not arrive until sometime after three o'clock. When we asked about the smoke, we were told that a big bush fire was burning about ninety kilometers to the south. Very little rain had fallen at Reliance during the last month, and further south they had had none. The bush had become tinder dry, and lightening had started the fire. There was no commercial timber in the area, and no access. The fire just burned. It had been burning for ten days, and would probably burn for another week, until moist arctic weather moved south. He told us that there was a strong continental weather system over the prairies that was pumping warm air north, and that we would probably have good weather for another week, perhaps longer, but after that we could expect an indefinite period of unstable weather that always accompanies the change of season.

Since we had about three hours free time, I decided to go looking for outcrops of the Pethei limestone, so we could collect stromatolite specimens. The nearest outcrops were about fifteen kilometers away, in the vicinity of Meridian Lake. I asked Fred if we had enough gas to get us there and back. He checked the tanks, spanned the distance on the map, and said we had enough with a bit to spare. We had brought only enough gas for a one-way trip (with a safety reserve), because Keith was bringing gas to get us back to base camp.

As we approached Meridian Lake, we had no trouble finding the Pethei limestone. It formed bare ridges, because the glacier had scraped the outcrops clean. A few solitary trees and an occasional open grove were the only obstacles to landing. Fred wanted to land on a high point, because, on this hot day, the south wind coming up the back slope of the ridge would give the helicopter extra lift during landing and takeoff. We found an open space near the top of the ridge and landed on its back slope. We searched the area for twenty minutes, and found only a few poorly preserved stromatolites. It was not a good collecting site.

When we returned to the chopper, I located the next large area of outcrops, where a different bed might be exposed, and asked Fred if we had enough gas to get there. We were nearly down to our safety reserve. Fred, however, allowed that we could get there, but no further. The second outcrop area was on a ridge that was as well-exposed as the first. Its back slope was made up of brown weathering limestone that formed benches up to a meter high. They dipped gently to the south. The benches formed a series of shallow steps, like seats in a Roman amphitheater. They ended at the shore of a small lake, about two kilometers away. Widely spaced spruce trees uniformly covered this long slope, except near the top where there were many open spaces underlain by limestone pavement. Fred found a landing spot on the pavement near the crest of the ridge.

As soon as the engine was cut, I scrambled down the back slope looking at the limestone. Fred followed as soon as the rotors stopped turning. I did not take my hammer, because breaking rock would not expose algal structures. The geology hammer is a crude instrument at best, and wholly unfitted for working with fossils. Freshly broken rock shows very little of the filmy lines of algal mats, domes, or cones.

Stromatolites are best exposed by the solution weathering of water that is slightly charged with carbonic acid from decaying vegetation. The weathering process preferentially dissolves the lime, and leaves the silica. The result is a finely detailed etching of the intricate patterns produced by the alternating layers of lime and silica, or lime and sand grains in stromatolites. It defines the finest internal structures. Frost action is the second most important agent for exposing stromatolites. It breaks the limestone into square blocks, or flagstones, that expose algal structures in three dimensions. Over many years, the weathering process has freed stromatolites from their matrix and allowed them to be studied in detail.

Our ridge was equally composed of bare limestone pavement, and a rubble of frost-cleaved flagstones. All of the exposed beds contained stromatolites, and a couple contained abundant remains of algal mats. About three benches down from the landing site, I found a bed that was wholly composed of stromatolites. For the next forty minutes, we wandered over the beds making piles of good specimens. Two steps beyond our first discoveries, we collected better ones, so we went back and scattered the inferior ones. This happened half-a-dozen times, until we had five large piles of select specimens. Some of them would make excellent cabinet specimens, and a few were good enough for museum exhibits. When it was time to leave, we made a quick survey of two lower benches to see if we had overlooked a better source of specimens, but we found nothing as good as the ones we had collected. We loaded our carrying bags, and slowly retraced our route to the chopper. By the time we reached it, we were carrying twenty-five kilos apiece.

When we had deposited our loads, I noticed that the white limestone beneath the chopper was full of algal structures that I had not seen when I got out of the chopper. I noticed them now because I had walked back to the chopper with my body bent forward and my face looking down, in order to help balance the load on my back. I spent a hurried fifteen minutes examining a short section of this outcrop, and gathering another ten kilos of specimens.

In the five minutes before we had to takeoff, I did a quick sorting job, but there were still over thirty-five kilos of specimens loaded on the chopper's carrying racks. We had plenty of lift for takeoff, and Fred flew along the backslope of the ridge just behind its crest in order to use the lift from the upward moving flow of air. We were back at Reliance at three o'clock. As soon as we got on the ground, I did another sorting job, and reduced the load to ten kilos, which was the maximum amount we could carry with full tanks, and another person aboard. When Keith arrived at 3:30 p.m., we unloaded the drums from the plane, gassed up, and started home.

Fred used the strong south wind to get us out of the Great Slave basin with a minimum expenditure of fuel. From Reliance, we flew east along the length of Fairchild Peninsula. He flew just above the ridge's crest, on its backslope, in order to get maximum lift from the south wind. We flew to where the Lockhart River enters the extreme eastern end of the Great Slave basin. By flying up the north side of Lockhart valley, Fred caught the updraft of the wind as it rose to escape the basin. We rode this updraft until the valley began narrowing into a canyon. At this point, there could be dangerous turbulence, so Fred turned south in order to fly up a side valley. At the same time, he applied full power and flew directly into the full strength of the south wind. The air speed indicator continued to register 110 kilometers per hour, but we slowed to less than thirty-five, until we climbed over the rim. Once on top, we rapidly accelerated. In two kilometers, we were at our prospecting altitude and beyond sight of the basin. In another ten kilometers, we were back in the barren lands.

CHAPTER 17

A Fossilized Microenvironment

During the week following our trip to Reliance, cool and misty weather gradually moved south, but the weather remained flyable. On several days, we logged seven hours flying time, and were absent from base camp for eleven or twelve hours, due to frequent stops to examine small rusty spots that could have indicated gold mineralization. On the morning of September 7th, a dripping fog blanketed our camp, and it did not appear that we could fly. However, in the previous afternoon, Shaver had flown to Yellowknife for a load of aviation gas. He got an early morning start on the return flight, because the weather report from Reliance was good.

He arrived at ten o'clock in the morning and told us that the front was stationary about fifteen kilometers south of camp. From Reliance he had flown parallel to it in bright sunshine, but when he got near camp he ducked under it and flew north, down the Thelon valley until he found camp. Shaver said the chopper would have no navigational problems, so Keith suggested we run a traverse to the south. We left our flight plan and departed within twenty minutes.

Sometimes groves of trees of considerable size grow a long way north of the treeline. I will describe three occurrences of this phenomenon: valley groves, esker groves, and a fossilized micro-environment. Our Thelon River base camp was located near several valley groves. Although the Thelon River is mapped as being completely north of the treeline, there are full size trees, (up to seven meters tall), growing in its valley, forty kilometers north of base camp. At its headwaters, the Thelon River flows eastward from several large lakes, on a high plateau in the barren lands. Then it turns north. After making the north turn, full size trees become common. This anomaly occurs because the river, while still 1,400 kilometers from its mouth, rapidly drops from approximately 450 meters to only 165 meters above ocean level. At this lower elevation, it flows through an area of well-drained sand and gravel terraces, which are prime sites for tree growth in the region of the treeline. Our base camp was located in the north-south section.

In the winter, shallow declivities in the valley are filled with drifted snow that insulates seedlings from the worst effects of winter. In the spring, percolating meltwater thaws the underlying sand while snow still covers the trees. The greenhouse effect of the snow cover produces optimum growing conditions early in the summer. Lower elevations also mean there is a thicker layer of atmosphere to convert sunlight into heat, and when this is combined with the protection of snow drifts that provide winter insulation on well-drained growing sites, small trees grow in the Thelon valley as far north as Eyeberry Lake.

The second micro-environment is associated with the eskers. In the region of our Thelon River camp, all eskers run east-west, and are composed almost wholly of sand. The prevailing north winds have shaped them into dunes, some of which are crescent shaped. Spruce trees grow on the southern slopes, in the centers of crescent shaped dunes. This environment is cognate to the well-drained, protected sites in the Thelon valley. Even 130 kilometers north of the treeline, at these isolated upland sites, trees grow up to seven meters tall, even when the next highest vegetation is only ankle high. These trees are often half buried by drifting sand, but they survive as long as they are in the lea of a dune, and their root systems remain in sand that thaws during the summer.

Valley and esker groves describe the normal way tree growth advances into the barrens, but sometimes this sequence is reversed. A reversal produces a fossilized micro-environment. An environment becomes fossilized when harsher climatic conditions prevent the germination of seeds, but do not kill mature trees. Until the mature trees die, they are reminders that slightly milder growing conditions once prevailed.

On our September 7th traverse, we found a fossilized micro-environment where the Elk River joins the Thelon River. The elevation was between 400 and 425 meters, and was about thirty kilometers north of the treeline. It was in the transition zone, where slightly milder climatic conditions could push the treeline many kilometers northward, or slightly harsher conditions could force it south an equal number of kilometers.

In this region, the hills and ridges are rounded and monotonously alike, and the valleys are wide and flat. The centers of these valleys have slightly incised water courses. Relief between the main water courses, and the highest hills is only ten to fifteen meters. A carpet of scraggly sedges and a few tufts of grass cover the hilltops. Groves of waist high arctic willow grow along the margins of shallow valleys, but the startling feature of these willow groves is that many of them have trees growing in them that are ten to twelve meters high. The trees are tamaracks, not black spruce. Many willow patches contain a solitary living giant, as well as many trunks slowly rotting on the ground—a reminder that a grove of mature tamaracks once grew on that site.

All the willow patches, and all of the tamarack groves, whether made up of bleached trunks or fifty living trees, are growing in shallow hollows along the sides of wide valleys. They do not grow in the lowest parts of the valley. The largest groves have a dense willow undergrowth, while there are several small tamarack groves that have no undergrowth, only fully mature and perfectly proportioned trees growing in the barrens. The amazing feature is their size. They are respectable trees. These tamaracks were taller than the trees growing 100 kilometers south of the treeline. Their height was accentuated, because there were no seedlings. There were only tall tamaracks and arctic willow bushes. Under normal treeline conditions, black spruce is the tree that grows furthest north, while tamaracks generally grow south of them, and willows grow north of both. But, there were no spruce trees growing among the tamaracks, not even dwarf trees.

Why were they growing there? The presence of tamaracks meant that the environment had once been very wet, because tamaracks usually grow in swamps. The seeds of the giant trees had germinated along the margins of swampy bays, and it was probably so swampy that spruce trees could not grow in the marginal conditions at the treeline. This environment agrees with the habitat for tamaracks growing south of the treeline. Transplant a swampy microenvironment to the treeline and tamaracks are a more competitive species of tree than black spruce.

Two hundred or 300 years ago, this region was covered by an extensive system of shallow, interconnected lakes, some of them large and all of them having numerous shallow bays with swampy shorelines. The lake system was more than 500 square kilometers in area, and probably covered 70 percent of the land surface. A nearly identical environment can be seen sixty-five kilometers to the west, at Eileen Lake, where the treeline bulges into the barren lands around the shore of the lake.

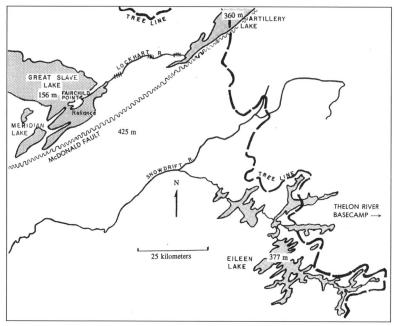

The treeline around the east end of Great Slave Lake.

In the valleys where tamaracks survive, all of the interconnecting lakes have drained. Drainage was a step-by-step process, as lake outlets were lowered by down cutting streams. After each drainage event, lake levels stabilized for several tens of years, and then drainage recommenced when erosion further lowered outlets. Many sand deltas and beachlines are present at lower elevations than those of the tamarack groves. Dwarf spruce trees, up to a meter tall, grow on them. These trees are recent colonizers compared to the groves where giant tamaracks survive.

This situation is the reverse of the normal colonization sequence. The reversal is due to the disappearance of a large heat reservoir in the form of a system of shallow lakes. During the summer, large amounts of heat were stored in shallow lakes by twenty-four hours of sunshine, and it especially warmed the water of shallow bays. This heat was released from the lake during autumn. In late August, just before the advent of winter, the shallow bays rapidly released the heat stored in their waters, which was a sufficiently powerful moderating influence to lengthen the growing season by two weeks along swampy shorelines. The extra heat, from the lake's water, was the reason why the tamarack seeds could germinate, and the insulation of snow drifts created conditions where growth could resume one or two weeks earlier than normal in the spring.

In summary, the conjunction of three geographic features produced conditions favorable for tamarack growth: 1) an extensive system of interconnected lakes with swampy shores; 2) shallow bays with shoreline swamps that delayed autumn freezing, plus enough runoff water in the spring to add one or two weeks to the growing season; and 3) drifting snow that filled the shallow valleys and insulated tamarack seedlings from desiccating winds and, at the same time, extended their growing seasons long enough for them to survive.

The system of shallow, interconnected lakes drained because erosion successively breached the natural dams that impounded water in the lakes. In the spring, when the outlet river was swollen with runoff water, it down-cut its channel, and a shallow lake drained. As soon as the level of one lake dropped, next spring's runoff water would find a sufficiently increased gradient to lower the outlet of another lake, and it would drain. A stable lake level would remain for a few years, until another dam was breached and another lake would drain. Once down-cutting began, drainage proceeded very quickly. In a few decades, the whole system of lakes and bays was reduced to a few small lakes occupying the deepest portions of former basins.

The vastly reduced surface area of the lakes, and the reduced volume of water, ended the moderating influence of the heat stored in lake water. The contraction of a lake system, to less than 5 percent of its former size, shortened the growing season and forced the treeline to retreat. The tamaracks that had seeded and grown to maturity on the shorelines of swampy bays were stranded. However, because they had reached maturity before the lakes drained, they survived. The growing season, though, was sufficiently shortened so that tamarack seeds could not germinate. Under newer, harsher growing conditions, the only seeds that could germinate were black spruce. They grow as knee-high bushes on the sand beach lines that mark temporary pauses in the draining of the shallow lakes. These beach lines are at lower elevations than the tamarack groves.

CHAPTER 18

Discovery

The Contwoyto Lake gold prospect was not discovered from the air. It was discovered by one of the ground prospecting teams operating out of a fly-camp. It is doubtful that the gold mineralization would have been seen from the air. The gossan was patchy, and its color was dark maroon. It would have been seen only if the chopper had flown directly over it, and there was no glare. Gossans capping gold mineralization in Precambrian rocks are almost always dark maroon, unlike the bright red gossans that usually cap base metal deposits.

On the basis of favorable rock types, Keith had located both prospecting parties on small lakes about ten kilometers west of the west shore of Contwoyto Lake. Contwoyto Lake is 120 kilometers long, its surface is at an elevation of 450 meters, and its long axis is northwest-southeast. It occupies an irregularly shaped basin on the height of land between the Back River, which drains from its southern end, and the Burnside River that drains from its northern end. The large size of the lake meant that the prospectors could traverse to the east from their fly-camp until they reached the west shore of Contwoyto Lake. When they reached the lake, they would know exactly where they were and this would minimize the chances of getting lost. This distance also set a limit on a day's prospecting traverse. Keith did not route helicopter traverses in the area where the discovery was made because ground prospecting crews were there to do this job, and there was an immense area of other land available for aerial prospecting.

When we moved base camp from Benchmark Lake to the Thelon River, Keith brought the two prospecting teams to base camp to provide labor for the move. This was standard procedure. On their last traverse before rejoining the Benchmark Lake base camp (August 26th), one prospecting team made a traverse that ended on a high, rounded hill overlooking Contwoyto Lake. On top of it, within 300 meters of the west shore, they found a gossan with highly favorable mineralization. They collected three assay samples and four hand specimens.

All of the assay samples and rock specimens contained rusty quartz, iron sulfides, and arsenopyrite. All three of these minerals are associated with gold mineralization. The next plane to Yellowknife carried the assay samples and specimens to our expediter, who airmailed them to INCO's chief geologist. Keith received the assay results on September 2nd's evening radio schedule. They were encouraging, even though a very high gold assa, at 1960 prices (thirty-five dollars per ounce), was required to undertake gold mining in the Arctic.

The favorable assays required an immediate decision. Should claims be staked, without knowing the size of the outcrop, and during the change of season when a staking crew could be confined to their tents for a week or more if there was a gale? Keith decided not to stake the claims so late in the season. Staking and investigation were postponed until the next field season.

After the Thelon River base camp was operational, both prospecting teams were flown to fly-camps where they prospected until September 6th. On September 7th, Keith had the plane fly them directly to Yellowknife. Upon arrival at Wardair's dock, they went directly to INCO'S warehouse where they changed into civilian clothing. With no time to spare, they were taken to the airport, where they boarded the plane to fly south to Edmonton. They were not going to stay overnight in Yellowknife, and get drunk and talk about their gold discovery on the shore of Contwoyto Lake.

CHAPTER 19

Snap Freeze

We stayed in the Arctic to the end of the season. Normally, all exploration ends in the first week of September, and all personnel and equipment are back in Yellowknife by the 3rd or 4th. We stayed longer, because we were near enough to Reliance so that, if there was a snap freeze, personnel and necessary equipment could be ferried there by helicopter, and then flown to Yellowknife by float plane. During the change of season, when weather on the upland plateau can be violent, float planes can usually reach Reliance by flying just above the surface of the big lake.

In deciding to establish the Thelon River camp, Keith gambled that flyable weather would continue until the middle of the month. Sometimes it does and sometimes it does not. We lost our gamble on September 11th. Change of season weather can be dramatic. Freezeup can descend on the land in one continuous blast that can stop most bush flying for two weeks or more. We knew summer was over because we had been flying in unstable weather for four days; but Keith gambled against an instant winter. For a day or two, there was clear weather with southern winds driving alto-cumulous clouds northwards. Then, the winds would reverse for one or two days and spread low clouds over the land. Weather like this might make a couple of cyclic alterations, but each time the northern winds were a bit stronger and cooler, and each night was fifteen minutes longer. Then, the weather would break. Winter would arrive with continuous gale winds, low overcast, sleet, ice fog, and snow flurries, until cold settled on the land. Thereafter, the weather would be crystal clear and dry with only occasional high winds during blizzards. If change of season weather is protracted, it produces a snap freeze.

We were almost caught. Change of season weather began on the morning of September 11th, and ended the evening of the 13th. It was not quite a snap freeze, but only because we were camped on a lake that was part of the Thelon River, which brought in a continuous supply of warmer water, and because we were at a lower elevation than the barren land plateau. We were camped at the best possible place for this time of year, because our lake would be one of the last small lakes to freeze. When we left, all of the small lakes on the barren land plateau had rims of ice on their shorelines.

Early in the morning of the 11th, the winds increased to an estimated sixty kilometers and by noon they were gusting to 100 kph. Our tents, which were nailed to plywood platforms, tilted ten degrees to the south, and the sides bulged inward so that we could lean against them like a cushion. When we went from tent to tent, we had to put on coats to prevent being chilled. To maintain our balance, we had to lean head down into the wind, so that our torsos were nearly parallel to the ground.

On the morning the 12th, when we went to the cook tent, there was a dusting of snow on the ground. Tufts of grass on the beach were ice-covered, and frozen into brittle, grotesque shapes. The chill factor was many degrees below freezing, and a person without heat and shelter would quickly die of exposure. The velocity of these winds was confirmed by our radio operator in Yellowknife, who reported that the weather station at Contwoyto Lake was experiencing similar weather. The winds were continuous for three days, never falling below 60 kph.

The ears of the two aircraft mechanics were attuned to the wind. On the morning of the 11th, they got out of bed when it was still dark, and weighted each of the chopper's carrying racks with three, full, ten gallon gas drums. Then, they filled the airplane's floats with water, and added a quart of anti-freeze to each compartment. After breakfast, we rolled full gas drums under each wing, as we had done during the breakup wind, and looped ropes over the wings and tied them securely to the drums. By late evening, the temperature was minus two degrees, and all of the rudder cables on the floats, and tie lines on the drums, were encased in ice.

All camp personnel were apprehensive because there was nothing to do but sit in our tents with space heaters producing their maximum heat. On the morning of the 13th, two to three centimeters of ice were on the land portion of the shoreline from spray hitting the cold ground and freezing. We knew that, if the wind continued for three or four days we would be leaving by helicopter. This would mean that all of our personal belongings would be left behind, to be retrieved sometime in the indefinite future, hopefully before freeze-up. Our nerves were on edge from the constant howl of the wind, and the tents that could not be kept warm. We had nothing to do but eat, drink coffee, and wait. All available literature was read by everyone, regardless of its quality, from the Brothers Karamazov to copies of Nugget and Stagg magazines. We all said an unexpressed prayer for a couple of days of flyable weather.

On the morning of the 14th, the wind lessened to 50 kph, but the ceiling began to lower. We could be gone with all of our equipment by the evening of the next day if the wind decreased to thirty kilometers per hour, and the ceiling didn't drop to the deck. At three o'clock in the afternoon, the wind abated to an estimated 25 kph, and the blanket of overcast broke into scattered clouds. The Beaver mechanic had already shaken the ice off of the rudder cables, drained the floats, and checked the engine to see that no sand had been blown under the cowling, or ice formed over the electrical system. The weather was flyable and Keith was going to try to move camp directly to Yellowknife.

The first load was aboard the Beaver by 3:30 p.m., which meant that Shaver would reach the big lake at dusk, and arrive at Yellowknife in the dark. There would be no navigation problem once he got to Great Slave Lake, because the reflection from the water would provide easy nighttime navigation if the weather remained reasonably clear. After depositing the load in Yellowknife, he would refuel and, weather permitting, return that night to Reliance where he would sleep at the weather station. He would arrive at camp an hour after dawn on the 15th.

While the Beaver was being prepared to leave, the chopper was also being readied to fly. The chopper would carry the chopper mechanic as a passenger, plus his tools, and personal belongings. It would be a full load, but they would be carrying a little extra fuel. They would fly to Reliance and remain there until tomorrow evening, in case the winds resumed and the remaining personnel had to be flown out by chopper. If camp was broken without further weather delays, the chopper would fly to Resolution on the south of Slave Lake, then to Hay River, and follow the road south to Edmonton.

The first plane load out of camp contained our geophysical equipment, exploration records, assay samples, a major portion of our personal belongings, plus Keith, who would inventory equipment as it arrived. He would also charter a plane for the following day, so we could break camp in one day. If a charter was available, we could move everything in three more planeloads.

The labor for breaking camp would be done by myself, Pete Dankas, and the Beaver mechanic. The cook, who remained with us, would have his hands full preparing meals, and getting his cooking utensils ready for moving. After the Beaver and chopper left on their twilight flights, we began taking down one of the tents, and carrying it to the beach so a load would be available when Shaver returned at dawn. We kept the radio in operation for scheduled communication in the morning, or until Shaver arrived, whichever happened first. Only then would it be dismantled, and put in the morning load.

Shaver returned at dawn with the news that a chartered plane would arrive at ten o'clock. This meant we would be able to fly out most of our heavy equipment, provided the weather held, but it also meant that Shaver and I would be working to the point of exhaustion. He would make two more round trips to Yellowknife that day, and I would leave on the last flight.

We loaded the chartered plane without incident, but it did not get away until after noon because we did not have a full load on the beach when it arrived. The canoe and Pete flew out in it. Shaver took his mechanic on the next trip, and I had six hours to lug the last equipment to the beach and stack the plywood floorboards and put the remaining drums of aviation gas and heating oil on top of them, to prevent winter gales from scattering them on lake ice.

I had a spasm of apprehension when I was alone on the beach. The weather might deteriorate again and leave me temporarily stranded under very unfavorable conditions. I was primarily worried that the wind would increase, and my senses were attuned to this, so I failed to notice that the ceiling gradually lowered. But, I had little time to worry, even if I had noticed this, because I had to work at full speed in order to accomplish what had to be done. I also had to keep working in order to keep warm.

Shaver returned at four o'clock that afternoon, but he came from the south instead of from the west. Then, I noticed the weather. There were sixty or seventy meters of ceiling over the lake, and much less over the lakeshore bluffs. Shaver had arrived by slipping through an opening in the overcast, and dropping into the Thelon valley. Once in the valley, he flew north to camp. It took us twenty minutes to load the plane and refuel. Shaver was extremely anxious to leave, because the overcast was just above freezing, and a drop of two or three degrees would produce icing. Such conditions would prevent flying, as would the fact that daylight would be gone in two hours. By the time we were airborne, a mist had settled on the bluffs around the lake.

It is very difficult to navigate in mist. One can see only the larger lakes. The pilot must know where he is at all times, or he will be lost, because it is impossible to fly higher and locate your position. Once you are lost in a mist or fog, you are permanently lost until it disperses. The only other option is to go high and fly by compass, and hope that Yellowknife is open. This is semi-suicidal at this time of the year.

After forty minutes flying, we were only 100 kilometers west of camp, and the overcast seemed to be settling to the ground. We were over Whitefish Lake. We knew exactly where we were, but we had nowhere to go. We sat down along the north shore, so that the north wind would blow us the length of the lake.

As soon as we were on the water and the engine was turned off, a couple of caribou appeared near the shore. We had seen several in the Thelon valley, so we were not surprised. Soon, four more were silhouetted on the skyline staring at us. When they were satisfied we were not predators, they began grazing. Eventually, they wandered down the shoreline and out of sight. Twenty minutes later, there was about 100 meters of ceiling over the lake.

We skimmed off the water, and turned south, down the length of the lake. We flew about half its length, when Shaver found a hole to the west. We had a ceiling of about 150 meters, and visibility was four or five kilometers. There were enough big lakes along our flight line so that we knew exactly where we were. Shaver flew with one-quarter flaps, and an airspeed of 150 kilometers, to give our loaded plane extra lift. We flew about 100 meters above the ground, and about twenty meters below the overcast. Occasionally, Shaver had to bank with the topography to avoid shrouded hills, and there was barely enough visibility to see over the low hills that separated the lakes.

Directly under us, the ground was distinct in every detail, but as the eyes were raised we often saw only the shrouded tops of lakeside bluffs. Because we were so close to the ground, we saw many groups of grazing caribou on their way to winter grazing south of the treeline. Unless they moved, they were extremely difficult to see. Their mottled brown and whitish coats blended with the muted brownish-green landscape under the overcast so that, in spite of the land having no vegetation more than ankle high, they were invisible when they lay down and remained motionless.

We flew a little north of west, gradually gaining altitude, but at the beginning of dusk our visibility nearly ran out. As Shaver was beginning to make a U-turn, to set down on a lake to spend the night, I spotted a corridor of visibility to the southwest. I tapped his arm and pointed. He saw it and straightened from the turn, and began threading his way through. Visibility was minimal, but it imperceptibly improved, opening like a needle valve into a funnel. We were rapidly losing visibility in the early dusk when, suddenly, we flew into open sky. Eighty kilometers further, in deep dusk, we were over Great Slave Lake, and safely on our way to Yellowknife.

We got up late, had a long breakfast, and began sorting equipment. We worked until early afternoon, and quit at three o'clock for beer. On the way to the Yellowknife Hotel we stopped at D.O.T. to get the weather at Reliance. The winds were up to 50 kph, visibility was down to a kilometer, the ceiling varied from 0 to 150 meters, and the temperature was one degree. Up on the barrens, conditions would be much more rigorous.

CHAPTER 20

Evaluation

What follows is an account of the next field season, in 1961, when I did all of the initial sampling on the gold bearing outcrops at Contwoyto Lake. We could not make a preliminary evaluation until the ice went off of the lake so that a camp could be supplied by float planes. Breakup did not occur until July 19, 1961. I was on the first plane when it arrived on July 22nd. We would be on the prospect for a week to ten days to make a preliminary evaluation.

We used three procedures to make a quick evaluation. The first was staking claims. Initially, two groups, totaling thirty-six claims, were staked to cover the prospect, and the immediate vicinity. The second step was a pace and compass map of the mineralized zone, and the third was collecting grab samples from mineralized outcrops.

Mapping was by sighting a compass line along the length of the mineralized zone. This formed a baseline. I walked the baseline, counted my paces, and noted the locations of all outcrops. In the evening, I drew the baseline on paper, and converted my paces into meters. On the measured baseline, I drew the locations of all outcrops. The next step was using the compass to turn off crosslines at right angles to the baseline. While walking the crosslines, I noted the locations of all outcrops and, while pacing both the baseline and crosslines, I collected rock samples (grab samples) for assay.

From the moment I saw the mineralization, I recognized that it was similar to the sulfide iron formation being mined in the Homestake Mine, at Lead, South Dakota, which I had visited as an undergraduate student. The Homestake Mine is one of the largest and richest gold mines in the world. It has been in continuous production since 1879. Gold bearing, sulfide iron formations consist of thin alternating beds of slica, iron silicate minerals, magnetite, and iron sulfide minerals (pyrrhotite and arsenopyrite), with the gold in the sulfide beds. Where there are no sulfide minerals there is no gold.

I wanted the assay samples I collected to be representative but, at the same time, I wanted them to contain above average values. I wanted to establish the presence and persistence of gold mineralization, because, if values were high and found over a wide area, this would be the best incentive for INCO to undertake an intensive investigation. High gold assays from many grab samples would indicate that an intensive investigation was warranted.

On the basis of the assays from the first set of grab samples, Keith recommended an intensive investigation, and INCO's chief geologist agreed. The needed personnel arrived as fast as they could be transferred from other programs. The shortness of the remaining field season dictated that everybody in camp had to labor when labor was needed, regardless of their primary skills. When bull labor was needed, all persons labored except the cook.

The first step in the intensive investigation was to make a more accurate map on which to plot: 1) the geology; 2) the magnetic intensity of rocks obtained by a ground magnetometer survey; 3) a ground electromagnetic (EM) survey to detect any anomalies caused by sulfide minerals; 4) the locations of all outcrops that were sampled; and 5) the results of the assays.

The more accurate map would be based on a measured baseline, and measured crosslines. The measured baseline was along the length of the zone of mineralized outcrops, on top of my pace and compass baseline and crosslines. A steel tape was used to measure the length of all lines. Uniform directions were maintained by sighting along three wooden pickets having sharpened points, like gun sights. When three pickets lined up, the direction was constant. The measured baseline was two kilometers long. Eventually it would be five kilometers long, as geologic mapping and ground geophysical surveys were extended away from the discovery outcrop.

Crosslines were turned off the baseline at ninety degree angles, at thirty meter intervals. Crosslines extended for 450 meters in both directions, except where they ended at the shore of Contwoyto Lake. Wooden pickets were placed at thirty meter intervals along the baseline, and on all crosslines. Each picket was one and a half meters high, and had the meterage marked on it with a carpenters pencil.

The outcrop map was plotted on the measured grid, but it did not give a clear structural picture because there were not enough outcrops. Outcrops formed only 10 to 15 percent of the area on the discovery hill, but the obvious was obvious. Gold mineralization was confined to the outcrops that had thin beds of sulfide minerals. There were only two visible sulfide minerals in the beds: pyrrhotite (iron sulfide), and arsenopyrite (iron/arsenic sulfide). The rest of the iron formation consisted of beds of silicate minerals, beds of quartz (that were originally chert), and lenses of magnetite. Magnetite gave the gold bearing rocks a magnetic signature, and oxidation of the sulfide minerals made gold bearing outcrops instantly identifiable.

Magnetic readings were taken at ten meter intervals on the measured grid. When the operator found a magnetic high, he backed up five meters, took another reading, and continued to take readings at five meter intervals until the magnetic intensity dropped. The magnetic survey showed six linear highs with sharp peaks in them, and the assays from additional grab samples, collected on the measured grid, continued to be encouraging.

A more accurate map was needed, of the mineralized zone, on which to plot a more detailed magnetic survey and an EM survey. A surveyed grid was made. On this grid, the magnetometer operator took readings on ten meter squares over the entire outcrop area, and on a five meter grid where he found magnetic highs. He did this by taking a reading along a crossline, then going ten meters to his left to take a reading, and ten meters to his right to take another reading. The whole outcrop area had readings on the corners of ten meter squares, and on five meter squares where the magnetic intensity was high. Zones of high magnetics exactly correlated with high gold assays, and indicated that the gold bearing iron formation was in the shape of a tightly-folded, elongated Z.

The EM survey showed nothing. This indicated that the sulfide minerals containing the gold were in discontinuous pods. This result was half-expected, because pods are the usual configuration of sulfide minerals in tightly-folded and highly metamorphosed rocks. The high pressures and temperatures generated by folding and metamorphism fragmented the weak sulfide beds, and some of the ductile sulfide minerals flowed like toothpaste into zones of lower pressure at the noses of anticlines and synclines. These sulfide minerals formed pods that did not have enough lateral continuity to make a measurable EM anomaly.

Assays of samples collected from the pace and compass survey, and the measured grid, confirmed the persistence of values. Many samples were high grade but at this stage of evaluation, only thirty claims protected the prospect. Many more claims were needed because the prospect had the potential of being a large gold mine. There was also a good possibility that one or more similar deposits might be present in the immediate area because gold deposits are seldom found as single occurrences.

A five part program was immediately undertaken: 1) staking adjacent ground; 2) blasting trenches in the discovery outcrops; 3) diamond drilling underneath the highest grade outcrops; 4) building a permanent base camp, because an intensive evaluation program would take at least two more field seasons; and 5) prospecting for gold on claims distant from the discovery hill. Both drilling and trenching were to begin as soon as possible, but the most pressing need was protection and this required staking more claims.

INCO's chief geologist visited the prospect to see the extent of mineralization. When he arrived, he was sitting astride a bundle of wooden pickets that would be used as claim stakes. Federal laws governing staking in the territories limit eighteen claims in one block, and only one block per person can be staked within a sixteen kilometer radius. However, claim blocks in the name of other persons can be staked next to each other. Within a claim block, one stake can be used for the corners of four claims, but, where the claims of one block adjoin claims of another block, two stakes must be used. The most efficient use of wooden stakes, which were in limited supply, was blocks that were six claims long and three claims wide.

In the evening of the day the chief geologist arrived, he and Keith drew a staking map. They estimated the number of claims needed for adequate protection. It was large. In order to guide the staking, it was decided to make a north-south baseline the length of the area to be staked. The baseline was put in place before staking began. The chief geologist and Keith did it with the aid of the chopper, while we prepared the stakes for planting.

Each stake required a metal tag, and we began nailing them onto the stakes in sequential order according to the staking plan. At the same time, the following information was penciled on each stake: 1) the name and number of the claim within a group; 2) the time of staking; 3) who staked the claim; and 4) the name of the person who bought the staking license (not usually the person who staked the claim). We put the time of staking at fifteen minute intervals, which is about the right length of time to walk 450 meters, (the length of the side of a claim). Preparing 150 stakes was an eight man job that lasted four hours. After we labeled the stakes, we sorted them into piles in the order they would be planted.

The next morning, each staker was given a claim map, so that he could examine his stakes to see that he had the right number and that they were properly marked. They were then wired into bundles of twenty to twenty-four, and a rope was looped around the bundle so it could be slung on the shoulder. Each man carried one bundle. We had to walk about three kilometers before staking began, and the loads were so heavy that we had to bend forward while walking. Later, as we acquired more ground, the chopper flew staking parties to their starting points.

Stakes were planted the easiest way, not according to the information written on them. The times on the stakes indicated that each post was planted by one man walking around each claim, but we planted them by walking parallel lines. Seven men walking parallel lines (sometimes having to detour around lakes), could stake three claim groups of eighteen claims each, (a total of fifty-four claims), during a long working day. All available persons were employed in staking. Even INCO's chief geologist carried a load and staked a line.

When we reached our assigned starting place on the baseline, we took a compass bearing due west, at right angles to the baseline, and began pacing 450 meters. At about fifty meter intervals, we set slab rocks on end, or piled four or five smaller stones on top of one another. This made a claim line in the same way that trees are blazed when claims are staked in the bush. At the same time, we tried to see the men who were staking parallel lines, in order to keep claim lines as parallel as possible.

When 450 meters were reached, the stake corresponding to the map location was withdrawn from the bundle and planted. Then, the staker walked north and south, and hoped he could see the men staking to his right and left. If he saw where they had planted their stakes, he sighted in the north-south claim line and piled stones to mark it. If there was a hill between the stakes, he built a high cairn on the hill that could be seen from both corner stakes. Upended stones and cairns marked all claim lines, although, only two boundaries had been walked. I always marked the corners of claims with high cairns, up to two meters high, if frost heaved rock slabs were available. High cairns at the corners made claim lines much more visible than a series of upended stones.

After a corner stake was planted, it was sprayed with fluorescent orange paint, and two or three orange plastic streamers were tied around its top, so that it could be seen from a distance. Bright orange made it easy to backsight on claim corners and, thus, maintain a constant direction. After all claims were staked, a helicopter inspection indicated that most claims were remarkably square. Working in the barren lands helped!

We staked a total of 275 claims in six days of intensive labor. At the end of the season, an additional forty-five were staked along the shore of Contwoyto Lake to prevent rival companies from setting up exploration camps near the prospect. At a minimum, we staked sixty-five square kilometers.

Planting claim stakes was a big, physical job, but an equally big paper job had to be done before the claims could be recorded. Every claim required a form to be filled out in quintuplicate, and each claim group required another quintuplicate form. The forms were a very rough legal description that required a lot of repetitive typing. Many errors were made, and a lot of time-consuming proof reading was required to catch them. This was the thankless task of INCO's chief geologist while he was at Contwoyto, and of Keith on the days when he was not plotting magnetic readings and contouring them. Since the claims were staked in federal territory, they could be recorded at Ottawa as well as in Yellowknife. The law gave stakers ninety days between staking and recording.

By delaying recording as long as possible, and doing it in Ottawa, the chief geologist hoped to avert a staking rush, and thus reduce competition until the next field season. By having a base camp built, INCO crews would be the first ones in the field, and could begin a full-scale diamond drilling program while ice still covered the lake. Diamond drilling would define how much gold was in each ton, and the tonnage of gold bearing rocks available for mining.

After staking was completed, I was assigned to the trenching crew as soon as personnel arrived who could handle explosives. We began on the discovery outcrop that was also the center point of the surveyed grid. The first step in trenching was drilling holes to hold dynamite sticks. This was done with a lightweight, gas powered percussion drill. It was light enough to be carried to the discovery site in a backpack. A second man carried the gas, drill steel, fuses, and caps. Both men took turns drilling.

The purpose of trenching was to get a representative sample that was not enriched. Holes were drilled a meter apart, to a depth of 60 to 100 centimeters, and charged with two or three mini-sticks. The holes were inclined, so that the blast would partially clean the trench. When a trench was blown, I cleaned out the rubble to expose a clean, vertical face, (or the nearest thing to it). Cleaning a trench was hard work. It was usually dusty or muddy, and it was full of small, weathered fragments, and numerous shards of quartz. A shovel was useful, but hands were more efficient because they could scoop. There were usually several large rocks that had to be moved aside, and occasionally a partially separated slab had to be pried away from the outcrop to get a flat face.

I tried to locate trenches so that a clean face could be blasted from an existing scarp, but a high percentage of the outcrops were low domes or pavements that were barely emergent from the tundra. Trenches in these outcrops were sampled from the bottom. The most accurate samples were channels that were chiseled into the vertical face of a trench. Channel sampling is a slow, tedious process, and was done in all weather except rain. Normally, channel sampling is a job of unremitting dullness, but the search for visible gold lightened the labor. The channel samples from the largest outcrops showed sulfide mineralization over impressive widths, with numerous specs of visible gold.

While outcrops were being drilled for placement of explosives, I mapped the outcrops by measuring the angles of bedding. Most were vertical. I walked all of the crosslines and, if an outcrop was between the crosslines, I paced to it at right angles, and converted the distance into meters. In the week before the diamond drill arrived, the largest outcrops on the discovery hill were trenched, sampled, and plotted on the surveyed map. The locations of all previous assay were transferred to the surveyed grid.

After dinner, the day's mapping information and sample locations were transferred from my field map to the master map at base camp. Field information had to be immediately transferred to keep a running account of completed work. Transfer was necessary to prevent information from being lost, because my field map became smudged, crinkled, and semi-legible from field use. The completed geological map, with the assay results on it, could then be overlaid by the magnetic and EM surveys, and the results correlated. This aided locating diamond drill holes, so that the maximum amount of information could be gathered during the remaining days of the field season.

The diamond drill and its crew arrived during the last day of staking. Preparing the diamond drill for operation was a major undertaking. It had to be assembled on the beach, bolted to a sled, and two water pumps put on the sled as passengers. Then, it had to be winched up the hill to the discovery outcrop. Winching was done with the drill's motor and cable. Drillrods were driven into the ground, until they reached permafrost, and the cable was looped around the drill rod at ground level. Holding the drillrod upright gave enough purchase for the drill to be winched forty or fifty meters at a time. The discovery outcrops were about 300 meters from the shore of Contwoyto Lake, and about fifty meters above the lake.

Water to operate the drill required pumps and hoses. One pump was at the shoreline, another was halfway up the hill, and the third was at the drill. The chopper slung coils of water hose up the hillside so they could be strung and connected. Drillrods were carried up the hill, but the helicopter lifted fuel drums and bags of chloride. Lake water was circulated in the drill rods to flush the sludge created by cutting. Drill water had to be a strong sodium chloride solution, because, if water was used directly from the lake, it would freeze when it came into contact with the permafrost. Once the water froze, drillrods could not be recovered.

The saline solution operated in a closed circuit. When it flowed from the hole, it went into two steel drums. The cutting sludge settled to the bottom of the first drum and the solution overflowed to the second where it was recirculated to the drill bit. Nonetheless, there was continual spillage and, every time a new hole was started, a large part of the chloride solution had to be dumped to make the move. The drilling solution had to be continuously replaced, and this required an instant supply of chloride or else the drill could not operate.

The first hole was drilled at a thirty-five degree angle, under the discovery outcrop. It recovered sixty meters of core, and returned a long intersection of good looking mineralization with numerous specs of visible gold. The drill could not go deeper because its tripod legs were made of telescoped aluminum tubes. These legs were designed for use in the barren lands, where there were no trees that could be cut down and used to make tripod legs. The telescopic legs were easily transported by float plane, however, they were not rigid enough to bear the weight and strain of a longer string of drillrods.

The drill program was a great success. During the final month of the field season, five additional short holes were drilled. Assays from the core confirmed the high values of the surface samples, and the lateral continuity of mineralization. The result of our preliminary evaluation was that the Contwoyto Lake prospect had the possibility of becoming an operating mine when the price of gold increased.

In order to prepare for future work, a permanent camp had to be built and access by float planes had to be improved by building a dock. During the last two weeks of the field season (August 25th to September 6th), two-thirds of the camp's personnel worked on constructing five permanent buildings. They were a cookery, an office, two bunkhouses, and a warehouse.

The site of our tent camp was unsuitable for permanent buildings. Our tents had been pitched on a wide, low beach at the head of a shallow bay. It was a good temporary site, because it was dry and flat, and tents could be quickly erected after being unloaded from float planes. It was, however, unsuitable for permanent buildings, because it was too close to the lake. During spring runoff, a rise in water level would flood the flat, and ice floes could be blown onto it and destroy the buildings.

Building a dock was the first project, because unloading building materials from float planes would be slow and awkward without one. The location of the dock was dictated by geography. The lakebed under the dock had to be sand, in order to prevent airplane floats from being punctured. The site also had to be at a place where there was a straight channel to the dock, so that planes could avoid boulders protruding from the shallow bay. There was only one place where the dock could be built. It was built by the drill crew, using various sized planks and boards salvaged from tent platforms. The posts supporting the dock were made from worn out drill rods, flown in from the camp at Muskox.

Drillrods were driven deep into the bottom sand. They could be deeply driven, because there is a vertical interface between the unfrozen ground beneath lakes, and permafrosted ground inland from lake shores. INCO proved the vertical interface during a previous drilling program. A vertical hole was collared on lake ice with three meters of water beneath it. The drill was in unfrozen ground from the lake bottom to its end. Another vertical hole was collared on land, about thirty meters inland from the shore. It was in permafrost for its full length of 127 meters.

We did not want anyone in Yellowknife to know we were building a permanent camp because this would indicate that INCO had found a mineral deposit of economic importance. Purchasing building materials in Yellowknife, and trucking them to Wardair's dock for transportation north would be unmistakable evidence that something of major significance was happening in the barren lands. It was impossible to bring building material from Yellowknife without arousing suspicion.

Building material had to come from another source, and our Beaver could not do the job. The source was Prince Albert, Saskatchewan, and the material would be transported using chartered planes. An Otter was chartered from SGA (Saskatchewan Government Airways), for the final two weeks of the season, to carry building materials to Contwoyto. About 200 kilometers south of the Contwoyto prospect was Taurcanis Mine, a gold prospect in the process of intensive evaluation. It already had a shaft, and several large buildings. To aid exploration the Department of Transportation, in cooperation with the Department of Northern Affairs, had constructed a public landing strip.

INCO chartered Wardair's Bristol cargo plane (now sitting on top of a post at Yellowknife airport), and used it to fly building materials and fuel from Prince Albert to Taurcanis. Prince Albert is 1,100 kilometers south of Yellowknife and is not a mining town. When building materials arrived at Taurcanis, they were loaded onto a truck, moved to a lakeside dock, loaded onto the SGA Otter, and flown to the Contwoyto dock. From the dock, all materials were lugged uphill to the building site.

The building site was on high ground, about 200 meters from the lakeside campsite. It was a small, raised beach about seventy-five meters long and fifteen meters wide. It was level and its sandy composition kept it well-drained, even in the wettest weather. Unfortunately, the path from the dock had to cross a bog that was spongy in the driest weather, and mush when the weather turned wet. This impediment to easy walking could not be avoided.

Four buildings were laid out in a line. The first to be erected was the cookery. It was twice as large as the other buildings, being sixteen by forty feet. The plywood sheets used to make the walls and floors determined the size of the buildings. These sheets were four by eight feet. The drill foreman was in charge of construction because he had erected this type of building many times. The first step was dragging six by six ground sills from the beach and leveling them into the sand. Some digging had to be done, but never more than a few centimeters. Then, three-quarter inch plywood sheets were nailed to them. They formed the floor. The walls were built on the floor, with window and door openings in them. When a wall was completed, it was raised and braced until the roof was added. Standard size doors and windows arrived with glass already installed. As soon as the roof was built, the windows and doors were fitted into their openings and caulked.

The inside caulking was done with mattress stuffing jammed into cracks by hand, while a caulking gun was used on the outside. The building had to be made as air tight as possible against the fine, wind driven snow that would filter through the smallest cracks during the winter. If snow got into the building, it would produce a damp, uncomfortable building when it melted, and it would also increase the consumption of the very expensive fuel oil in the space heater.

The roof was made of shiplap boards that required a vast amount of nailing, but shiplap made a much sturdier roof than plywood sheeting. This extra strength was needed to resist winter winds that were strong enough to unroof and collapse fragile buildings. After the roof seams were caulked, heavy tarpaper was nailed on. The building was completed by lining the interior with fiberboard with good insulating qualities.

Extra labor to erect the buildings came from two sources: the chopper pilot and his mechanic, and the local Inuit. During days of unflyable weather, the chopper pilot did interior carpentry. He was good at it, because he had done most of the interior work in the house he lived in. The second source was two Inuit families living within long walking distance of the prospect. They subsisted on caribou and traded at Coppermine River although the Bathurst post was closer. The men of the two families were named Abraham and Andrew. Abraham was the less educated of the two. He knew almost no English. Andrew had acquired some knowledge of it while residing at a southern tuberculosis hospital, where one lung had been removed. While there, he had also acquired a taste for rum, and skill at cards. In the continuous after dinner game of rummy, (at twenty-five cents a point), played by the drill crew, cook, and others, Andy made a few dollars.

Andy had a son named John Peter who had spent some time in school at Coppermine River, and had learned some English. He had also learned to use a knife and fork. Abraham had three or four children, but, like their father, they had not left their home, and knew no English, nor how to use a knife and fork. However, when food was set before them, they learned very quickly and, in the rush to eat (we were eating in two shifts), no one took much notice of their fumbling.

Both families liked to visit, and Andrew liked to trade. He did a brisk trade in gloves, moccasins, white fox skins, and soapstone carvings that he had been taught to make at Coppermine River, where he purchased soapstone on credit. May, his wife, was a careful seamstress. They liked to visit as a family because we supplied free meals. The universal favorite was bread covered with butter and jam, washed down with tea and coffee, supplemented with large amounts of milk and sugar. It was hospitable to feed them, and they took full advantage of it, so that they became a mild nuisance. However, they never stayed longer than two days, and were always well-behaved. One day, in partial repayment, Andrew gave us a leg of caribou that made a nice lunch for the crew.

During the last week of the season, when the students started leaving, we ran short of bull labor and Keith decided to give Inuit labor a try. We sent the chopper after Andrew, who, as luck would have it, was not visiting our camp the day we needed him. He was anxious to earn money. He knew its value. Andrew was a willing worker, but had no stamina. The loads of material he carried from the beach were about half or two-thirds the size of our loads, and he could carry only two loads before he had to rest. He was not entrusted with any more skillful work, except nailing. He was quite competent at this, as was his son, who thought it great fun. At the end of two days, Andy had to go back to feed his dogs, or so he claimed. He did not like commercial labor.

We found out later that his wife was unhappy with his return. She had wanted him to earn as much money as possible, because she was quite capable of hunting caribou and feeding the dogs. In two days, he earned enough money to buy a pint of rum, and some other staples (tea, sugar, and tobacco). Just before we broke camp, he and his family returned to get their groceries. They stayed the night. He and May tapped the rum, got boisterously drunk, and made three-dimensional love in the tent next to mine. However, they were not so badly drunk that they were immobile in the morning. They did, however, have a hangdog look.

By this time, Abraham heard of the job opportunity. He walked to our camp and was hired. He was a less willing worker than Andy, or perhaps less skilled. He worked at about the same efficiency as Andrew, in spite of having both of his lungs. He too quit at the end of two days. Our Inuit labor experiment was over, and it was not wholly successful.

The last week of the season was a period of unstable weather. Most days were overcast with a dripping or misty rain, often with high winds, and the nights were freezing. Several nights it snowed a few millimeters, but it did not remain on the ground. The Peacock Hills, thirty kilometers to the north, however, had snow on their peaks for three days before melting. Fortunately, mapping was complete. The only continuing geologic work was diamond drilling, and trenching. After Keith logged the drill core, I bagged and labeled it and sampled the last trenches made in the smallest outcrops. Most of the time, however, I worked on building construction.

The last building to be erected was the warehouse. It was located at the head of the dock. It was built from materials left over from the other buildings, and lumber and plywood sheets salvaged from tent platforms. These materials became available after the students went south. Dismantling tent frames and platforms was time consuming, because all the nails had to be pulled. Fortunately, a canoe was available to carry the lumber and plywood to the dock, where the warehouse was being built. The canoe also carried the folded tents, bundles of aluminum tubes that supported the tents, stoves, and other field gear that would be stored in the warehouse until the next field season. The canoe was not INCO's. Four years previously, it had been left on the shore about a kilometer from camp by the Fish and Wildlife Service, when their personnel had tried to teach the Contwoyto Lake Inuit bands to fish, in order to have a more reliable food supply for themselves and their dogs.

During the two weeks the SGA Otter was freighting building materials from Taucanis, the Beaver made as few trips as possible to Yellowknife. It flew there only when camp was running low on groceries, or the mail had to be picked up, or a person from the home office had to be fetched. Normally, a trip to Yellowknife was made about every three days, but Keith tried to reduce it to once every five days.

In spite of this, INCO's activity attracted attention in Yellowknife, because it was known who was chartering the flights between Prince Albert and Taurcanis. On a trip north from Yellowknife at the end of August, our Beaver was followed by a plane flying high above its tail. Shaver was not aware of being followed until he was on the water at Contwoyto. The plane then made several low passes over the discovery site on the hill, where the drill tripod was outlined against the sky, and where there was a profusion of highly visible trenches. He saw the nine tents at the head of the bay and the permanent buildings in the process of construction. It was obvious that the project was substantial. To an informed observer, it was obvious that the mineral being sought was gold, because the trenches were not bright red and the gossans were smaller than a geologist would expect on a copper, nickel, or other base metal prospect.

After our activity was detected, we expected a rush of prospectors and stakers. Surprisingly, this did not happen. We had the area to ourselves for the remainder of the season. Why? It was known in Yellowknife that INCO did things big when they found potential economic mineralization. Local exploration companies could not compete against this. Yellowknife prospectors were willing to wait until INCO filed its claims, so that the limits of INCO's ground were known. Then, they could tie onto INCO's claims during the winter, when there was easy access with ski-equipped planes landing on frozen lakes. During that winter, at least 300 claims were staked on ground adjoining INCO's claim blocks.

Our final effort, after the warehouse was completed and the tents stored, was to stake claims along the shore of Contwoyto Lake.

The ground was underlain by granite, and did not contain anything of economic interest, but we wanted to tie up the ground so that a competing mining company would not have easy float plane access to interior ground. It was a precautionary measure to lessen the competition we anticipated next season.

We went to the staking area by the plane and canoe. The claims were staked in a sleet storm that turned to snow. Increased wind and freezing sleet in the afternoon grounded the plane, and darkness now came at 5:30 p.m. We were on our own to return to camp. The canoe could carry only half of the crew. Half were taken halfway, and put ashore. I was among them. We walked from there. The canoe returned for the rest. The entire crew arrived at almost the same time. All were soaking wet. However, the strenuous walking of my crew, who had been put ashore, kept us warm compared to those who returned in the canoe. They were deeply chilled, but hot coffee in a warm building prevented hypothermia.

On the next day, September 6th, I few south and the remaining personnel left the following day. On the way south, we flew over a swath of snow several kilometers wide and several centimeters deep. It would remain until spring runoff.

CHAPTER 21

Lupin Mine

The Lupin mine is the gold prospect on the shore of Contwoyto Lake that was discovered at the end of the 1960 field season, and evaluated during the 1961 field season. It is located 400 air kilometers northeast of Yellowknife, 100 kilometers in the barren lands (east of the treeline), and ninety kilometers south of the Arctic Circle. The first brick was poured on May 4, 1982, (twenty-two years after discovery), and commercial production was reached in October 1982. Two coincidental events made it possible to bring the Lupin mine into production. The first was the steep increase in the price of gold in 1979, and the second was the steep increase in the price of silver in 1980. Both metals had price spikes in 1980. For a few weeks silver sold for $50 per ounce, and gold for $959 per ounce. The following are year end closing prices.

Year	Gold ($/ounce*)	Year	Silver ($/ounce*)
1977	161	1977	4.84
1978	208	1978	4.93
1979	459	1979	6.25
1980	594	1980	12.53
1981	400	1981	8.43
1982	447	1982	5.57
1983	380	1983	8.83
1984	309	1984	6.69
1985	327	1985	5.88
1986	390	1986	5.36
1999	280	1999	5.13

*One troy ounce = thirty-one grams.

The Lupin mine is owned and operated by Echo Bay Mining Corp. It takes its name from a bay on the east shore of Great Bear Lake, 280 kilometers northwest of Lupin mine, where it mined a small, high grade silver-copper ore body. Echo Bay is five kilometers south of Port Radium, where Eldorado Mining Company operated the first uranium mine in the world. By 1977, the end of production at Echo Bay was in sight (March 1982), and the managers began looking for other investment opportunities. The steep increase in the price of silver in 1979 gave them the option of undertaking a large grassroots exploration program (a high risk gamble) or purchasing the Contwoyto Lake gold prospect and financing a diamond drilling program to define its tonnage and grade. Echo Bay Mines had money and managers who were experienced in operating a mine where no roads existed and where a large part of the supplies arrived by air.

In the early 1970s, Nelson B. Hunt, and his brother Herbert, began buying silver bullion and storing it in vaults in Swiss banks. The Hunt brothers were heirs of a Texas oil fortune. They purchased silver when its price was fluctuating between $1.30 and $2.00 an ounce. In the late 1970s, they used their money, and money provided by Saudi businessmen, to purchase silver futures. The brothers had a purpose. They wanted to corner (control) the world price of silver by holding more contracts for future delivery than there was newly mined silver or demonetized silver coins. This would force the principal industrial users (jewelry, photography, and electronic circuitry) to pay very high prices for the silver they owned or controlled.

By mid-1979, they owned 63 million ounces in their name, and 8 million additional ounces in a company they controlled. In January 1980, when the price of silver reached $35 an ounce, they got greedy and purchased future contracts for the delivery of 19 million additional ounces. The price of silver soared to $50 per ounce. In order to reduce price volatility caused by the Hunt brothers' speculation, COMEX (N.Y. Commodity Exchange) required all contracts to be actually delivered. On March 27, 1980, the price of silver plunged from $40 an ounce to $12. During this speculative frenzy, Echo Bay sold its future production forward for high prices. The result was that Echo Bay Mines had a bundle of cash to invest.

There was no single cause for the rapid increase in the price of gold in 1979 and 1980. Foreign and domestic events, however, created economic and political uncertainty in the United States and Europe. In November 1979, the clerical-political leaders of Iran authorized Muslim zealots to seize the US embassy in Tehran, and hold the staff hostage. The US imposed economic sanctions, and encouraged its allies in Europe to do the same. OPEC supported Iran (and its own interests) by raising oil prices 120 percent. On Christmas day in 1979, Russia invaded Afghanistan. The US response was an embargo on grain sales to Russia. Frigid political relations with Russia forced President Jimmy Carter to withdraw the SALT II treaty from ratification by the US senate. The treaty was designed to limit the nuclear weaponry of both nations.

During 1979, the annual rate of inflation in the US grew to 13 percent and, in 1980, it accelerated to 15.5 percent, and then to 20 percent. Inflation produced two results: the unemployment rate increased to 7.8 percent, and owners of US currency in the rest of the world began to sell it, which contributed to increasing the rate of inflation in the US. The status of the US dollar, as the preferred international currency, was under very strong negative pressure. Many people in the US, and elsewhere, looked for a safe place to protect their money from inflation. Many chose to purchase gold.

INCO was willing to sell the Contwoyto Lake prospect to a junior company that could pay cash and had expertise in arctic operations. Echo Bay acquired ownership in February 1979. The first step, to bring it into production, was definition drilling to confirm the tonnage and grade of the ore to a depth of 200 meters. Diamond drilling equipment, supplies, and personnel were brought in by float planes, the same method used by INCO during the initial evaluation in 1961. By mid-1980, proven reserves were 3.5 million tons, grading eleven and a half grams per ton, and indicated reserves were twice that amount with a similar grade. The decision was made to undertake mining.

The high price of gold between 1979 and 1982 made it comparatively easy to raise the 135 million dollars necessary to bring the mine into production. Lupin would be a medium size mine with a milling rate of 1,200 tons per day. When the decision was made to mine the Lupin orebody, its gold inventory was 2.5 million ounces. Four years after production began (1986), it cost about $172 to produce one ounce of gold that sold for an average of $375 per ounce. The Lupin mine was highly profitable.

After the production decision was made, it was necessary to build an airstrip, so that large cargo planes could bring in all construction materials. A 1,500 meter gravel landing strip was built, and later extended to 1,900 meters. All construction equipment and materials arrived in charter cargo planes, or in Echo Bay's Hercules

C130 cargo plane. A Convair passenger plane rotated construction personnel in and out of the minesite, and also carried high priority cargo. After the airstrip was operational, summer construction was twenty-four hours per day, seven days per week. It was necessary to get as much enclosed space as possible, so that machinery installation could continue during the winter.

If the mine was to be profitable, fuel and other operating supplies had to arrive overland. This required a winter road. In January 1983, a 667 kilometer road was completed between Yellowknife and Lupin. The road was relatively easy to build, because 530 kilometers was on lake ice, and much of the non-lake portion was in the barren lands, or through scrub forest that was easily flattened by a bulldozer. Convoys of trucks began using the road when lake ice was about a meter thick. It operated about twelve weeks each winter, beginning about the first of January. Round trips took three days, if the weather cooperated, which it usually does during the coldest part of the winter. One day was required for each leg of the trip, plus a day to unload.

Road maintenance is a problem during periods of high winds, because snowdrifts block land portions. Lake portions, however, are generally free of drifts. The road is kept serviceable by heavy-duty graders. Their high profiles allow them to clear the drifts that usually accumulate behind clumps of vegetation or in low spots in the topography; and where the road leaves lake ice. When the road is on lake ice, small trees are planted every 300 meters to mark the track. After a storm, a helicopter patrol locates drifted sections.

The completion of the winter road, coupled with ease of mining, made it possible for the mill to process more ore than originally planned. In 1985, additional machinery increased milling capacity to 1,650 tons per day, with no increase in the number of underground miners and mill personnel. Several factors made it possible to increase production without raising costs. The most important was permafrost.

Permafrost extends 530 meters below the mine surface. The entire production comes from permafrosted ground. Permafrost makes strong walls and roofs. Fractured rocks do not cave because fragments are frozen together. Therefore, little rockbolting and wire screening is required to keep tunnels and other work spaces open. This makes Lupin a very safe mine to work in. The only complication is lubricating the rods that drill the holes in which explosives are inserted. If water were used, it would freeze the drill rods in the permafrosted ground. Like the drilling done from the surface in 1961, underground drills are lubricated by a 7.5 percent salt solution, fed to the drills via rubber hoses, from sumps located near mining faces.

A second factor is ease of access. All broken ore is brought to the surface by a shaft, but the mine also has a corkscrew decline (spiral tunnel) that is large enough for all underground equipment to be brought to the surface for routine maintenance and repair. Equipment downtime is minimal.

A third factor is ease of gold recovery. Ease of recovery means that the gold is native, not chemically combined in minerals containing arsenic, antimony, bismuth, or tellurium, which makes recovery difficult and expensive.

A fourth factor is the consistent grade of the ore. There are no wide variations in grade over short underground distances, which require careful blending in order to feed a consistent grade through the mill. When broken ore reaches the mill, it is ground to fine sand-sized particles and held in leach tanks containing sodium cyanide. Cyanide dissolves gold. Milling practice retains the ore sand in leach tanks for thirty-five hours. During this time, 95 percent of the gold goes into solution. Wide variations in grade would prevent complete recovery, because ore containing excessive amounts of gold would not remain in the leach circuit long enough for all of it to go into solution. A predictable length of time in the leach circuit means that there is a predictable flow of ore from the mine through the recovery process; and that recovery rates are equally predictable.

Gold precipitates from the cyanide solution on a series of cloth filters that are impregnated with powdered zinc. Zinc replaces the gold. The gold drops out of solution, and is trapped on the cloth that contained the zinc. The gold impregnated cloth is taken to a furnace where it is burned and the gold melts. Melted gold is cast into bricks (dore) that contain about 85 percent gold, 12 percent silver, and 3 percent copper. The bricks are flown to the Canadian Mint where they are refined and cast into bullion bars of 99.99 percent purity. The fine sand, that remains after the gold has been recovered, is put into a slurry (with a little salt added in the winter to prevent freezing), and piped to a tailings pond where it becomes sediment at the bottom of a man made lake.

In 1989, Lupin's work force was 400 people, consisting of 200 who were working on site, and 200 who were on rotation. On-site personnel work for two weeks, seven days per week, in two twelve hour shifts per day. They then rotate home for two weeks. On-site workers are housed in two person bedrooms, in residential dormitories, that are connected to work places by covered walkways. The kitchen is always open to provide food. A high percentage of the work force lives in Edmonton, 1,400 kilometers to the south. Labor turnover is low. The men like a job that allows them to be home for two continuous weeks per month. The managers like the rotation system because the men concentrate their energies on doing their job. After completing their shifts, the men immediately eat, and, if inclined, play seven innings of softball in the summer or a period of hockey in the winter, or read in the library, or watch TV.

The instant profitability of the Lupin mine, and the continued high price of gold, allowed Echo Bay's management to use current and projected profits as a springboard for future growth. Growth depended on increasing gold reserves. In the 1980s, the best place to look for gold ore was in Nevada. Nevada was elephant country because newly developed heap leach technology, coupled with the availability of giant trucks built to haul copper and iron ore from open pit mines, made it profitable to haul low grade gold ore from open pit mines. These technologies gave value to gold bearing rocks that had been considered waste. Exploration activity by many companies indicated that rocks containing small amounts of gold existed in large quantities in Nevada.

Echo Bay's management made the correct decision. It was more profitable to mine large, low grade gold deposits in Nevada, than explore for smaller, higher grade deposits in Canada. Echo Bay's management saw an opportunity to acquire very large ore reserves in Nevada that companies with operating mines had missed. In January 1985, Echo Bay purchased Copper Range Mining Company for 55 million dollars cash. Copper Range owned a mine, mill, smelter, and refinery, in the upper peninsula of Michigan that had ceased operations in 1982. Echo Bay was not interested in reviving copper mining. The copper mine, and related infrastructure, was sold to a German mining company.

Echo Bay was interested in another asset. Copper Range owned 50 percent of the Round Mountain gold mine in Nevada. At the time

Echo Bay acquired half ownership, the grade of ore mined from its open pit was 1.3 grams per ton, (a little over one part per million). This is very low grade but there were 42 million tons of open pit ore where gold could be recovered by heap leaching. Heap leaching is a cheap technology and its efficiency was rapidly improving. There was also every indication that ore reserves were very much larger.

During 1985, aggressive exploration increased reserves to 175 million tons, and mining was increased to 17,700 tons per day. By 1997, ore reserves at Round Mountain were 200 million tons, grading 1.1 grams per ton, and the rate of mining had increased to 56,200 tons of ore per day, with a recovery rate of 75 percent. After twelve years of mining, ore reserves were larger than when Echo Bay acquired ownership. In 1997, production was 477,000 ounces and Echo Bay's share was 238,500 ounces. Gold was produced at a cost of $207 per ounce. Round Mountain is the biggest heap leach gold mine in the world. Acquiring half ownership of the Round Mountain mine was a spectacularly successful investment.

Early in 1997, the price of gold began a long slide that bottomed at $250 an ounce in 1999. Flexibility of operation was impossible at Lupin, because its location in an arctic desert requires it to be a self-contained operation. Isolation requires high fixed overhead costs to operate. To pay these costs, Lupin had to operate at optimum capacity. Production could not be increased because the mine and mill were already operating twenty-four hours a day, seven days a week, and less production would excessively increase overhead costs. Neither was it possible to increase the grade of ore processed by the mill.

Since commissioning in 1982, the average grade at Lupin had declined as the mine got deeper. In 1997, ore grade averaged 8.3 grams per ton. Lupin had 543,000 ounces of gold in proven tonnage of this grade. This tonnage was enough to last for three additional years of mining. In addition, probable ore, based on widely spaced drill holes, indicated that 500,000 additional ounces could be recovered at a profit, if the price of gold returned to average prices between 1986 and 1996.

The decline in the price of gold put enormous pressure on Echo Bay's management to reduce costs. In 1997, the company had four operating mines that produced 721,000 ounces (20.44 metric tons) of gold per year. Lupin contributed 165,000 ounces at a cost of $284 an ounce, but the mine was no longer profitable. Mining ceased in January 1998, after a total production of 2.8 million ounces (79.59 metric tons). Lupin mine was put on a standby basis for a year, meaning that the on site inventory of equipment and supplies was maintained. This cost 2.5 million dollars per year. In 1999, Lupin was brought back into production on a break-even basis in order to reduce standby maintenance costs to manageable amounts—in anticipation of higher gold prices.

Additional Reading

Bullis, H. Ralph., R. Andrew Hureau, and B. D. Penner. 1994. "Distribution of Gold and Sulfides at Lupin, Northwest Territories." *Economic Geology.* Vol. 89. No. 6.

Dingwall, Laima. 1985. "The Lupin: A Mine that Exceeds Even Its Own Expectations." *Canadian Mining Journal.* August. Vol. 106.

Dyke, Arthur S., Thomas F. Morris, and David E. C. Green. 1991."Postglacial Tectonic and Sea Level History of the Central Canadian Arctic." *Geological Survey of Canada.* Bulletin 397.

Fahrig, Walter F. 1987. "The Tectonic Setting of Continental Mafic Dyke Swarms: Failed Arm and Early Passive Margin." In Henry C. Halls, and Walter F. Fahrig, eds. *Mafic Dyke Swarms.* Geological Association of Canada. Special Paper 34.

Fraser, Hugh S. 1985. *The Great Thompson Nickel Discovery.* Thompson, Manitoba: INCO.

Gerhard, Lee C., William E. Harrison and Bernold M. Hanson, eds. 2001. *Geological Perspectives of Global Climate Change.* Tulsa, American Association of Petroleum Geologists.

Guthrie, R. Dale. 1990. *Frozen Fauna of the Mammoth Steppe: The Story of Blue Babe.* Chicago, University of Chicago Press.

Heaman, L. M., A. N. LeCheminant, and Robert R. Rainbird. 1992. "Nature and Timing of Franklin Igneous Events Canada: Implications for a Late Proterozoic Mantle Plume and the Break-Up of Laurentia." *Earth and Planetary Science Letters.* No. 109.

Hoffman, Paul F. 1988. "Pethei Reef Complex (1.9 Ga) Great Slave Lake, N.W.T." Memoir 13 in Helmut H. J. Geldsetzer, Noel P. James, Gordon E. Tebbutt, eds. *Reefs: Canada and Adjacent Areas.* Calgary, Alberta: Canadian Society of Petroleum Geologists.

Hoffman, Paul F., and Daniel P. Schrag. 2000. "Snowball Earth." *Scientific American.* January. Vol. 282. No. 1.

Hofmann, Hans J. 1981. "Precambrian Fossils in Canada: The 1970s in Retrospect." Paper 81-10 in F. H. A. Campbell, ed. *Proterozoic Basins of Canada.* Geological Survey of Canada.

Irvine, T. Neil. 1980. "Magmatic Infiltration Metasomatism, Double-Diffuse Fractional Crystallization, and Adcumulus Growth in the Muskox Intrusion and Other Layered Intrusions." In R. B. Hargraves, ed. *Physics of Magmatic Processes.* Princeton, Princeton University Press.

Khalil, Mohammad A. K. 2000. *Atmospheric Methane: Its Role in the Global Environment.* Berlin, Springer.

Kearey, Philip and Frederick J. Vine. 1990. *Global Tectonics.* Oxford, UK: Blackwell Scientific Publications.

Kerr, Daniel E. 1996. "Late Quaternary Sea Level History in the Paulstuk to Bathurst Inlet Area, Northwest Territories."*Canadian Journal of Earth Science*. Vol. 33.

Kerr, Daniel E. 1994. "Late Quaternary Stratigraphy and Depositional History of the Parry Peninsula-Perry River Area, District of Mackenzie, Northwest Territories." Geological Survey of Canada. Bulletin 465.

Kerswill, J. A. 1993. "Models for Iron-Formation-Hosted Gold Deposits." In Rod V. Kirkham, W. D. Sinclair, Ralph I. Thorpe, and J. M. Duke, eds. *Mineral Deposits Modeling*. Geological Association of Canada. Special Paper 40.

Kurten, Bjorn. 1979. *The Age of Mammals*. New York, Columbia University Press.

LeCheminant, A. N., and L. M. Heaman. 1989. "Mackenzie Igneous Events, Canada: Middle Proterozoic Hotspot Magmatism Associated with Ocean Opening." *Earth and Planetary Science Letters*. No. 96.

Lister, Adrian, and Paul Bahn. 1994. *Mammoths*. New York, Macmillan.

Nisbet, E. G. 1990. "The End of the Ice Age." *Canadian Journal of Earth Science*, Vol. 27.

Pielou, E. C. 1991. *After the Ice Age: The Return of Life to Glaciated North America*. Chicago, University of Chicago Press.

Pielou, E. C. 1994. *A Naturalist's Guide to the Arctic*. Chicago, University of Chicago Press.

Rowan, Roy. 1980. "A Talkfest with the Hunts." *Fortune Magazine*. August 11. Vol. 102.

Ryan, William B. F., and Walter C. Pitman. 1998. *Noah's Flood: The New Scientific Discoveries About the Event that Changed History*. New York, Simon and Schuster.

Sassos, Michael P. 1986. "Echo Bay's Lupin Mine: Where Highly-Mechanized Mining, Along with Modern Processing Facilities and Tailor-Made Solutions Win Arctic Gold." *Engineering and Mining Journal*. Vol. 187. No. 12.

Schopf, J. William. 1999. *Cradle of Life: The Discovery of Earth's Earliest Fossils*. Princeton, Princeton University Press.

Simpson, Sarah. 2000. "Melting Away." *Scientific American*. Januaray. Vol. 282. No. 1.

Tigray, Jeffrey H. 1982. *The Evolution of the Gilgamesh Epic*. Philadelphia, University of Pennsylvania Press.

Tully, Shawn. 1980. "A Hunt Crony Tells All." *Fortune Magazine*. June 30. Vol. 102.

Vatter, Cheryl L. 1989. "Operating Practices at Lupin Gold Mine, Cornerstone of Echo Bay Mines Ltd." *Mining Engineering*. April. Vol. 41.